24 Hour Telephone Renewals 020 8489 4560
HARINGEY L
THIS BOOK MUST BE RETU
THE LAST DATE M
To

Tim Radford joined the *New Zealand Herald* as a reporter aged sixteen and moved to the UK in 1961. He is a freelance journalist and a founding editor of Climate News Network. He worked for the *Guardian* for thirty-two years, becoming – among other things – letters editor, arts editor, literary editor and science editor. He won the Association of British Science Writers award for British Science Writer of the Year four times and a lifetime achievement award in 2005. He is an honorary Fellow of the British Science Association, and a Fellow of the Royal Geographical Society. He is the author of *The Crisis of Life on Earth: Our Legacy from the Second Millennium* and *The Address Book: Our Place in the Scheme of Things*.

70001731048 6

'Beautifully crafted "love letter to physics". . . His deft narrative interweaves discoveries such as the Higgs boson, the Hubble Deep Field and gravitational waves with Dante Alighieri's epic fourteenth-century poem *The Divine Comedy*, which intuited the laws of motion found by Galileo Galilei some 300 years later.'
Barbara Kiser, *Nature*

'Engaging and delightful . . . In Radford's persuasive and genial company, as he roams from the initial singularity to dark energy, from Saint Augustine's *City of God* to Dante's *The Divine Comedy*, from the Higgs bosun to the multiverse, it's hard not to be moved by the fact that there are those who are capable of dreaming up and executing complex undertakings that explore the order that underpins creation.'
Manjit Kumar, *Observer*

'Lyrical hymn to space exploration, knowledge and the enquiring mind. . . Helps quench our curiosity, yet deepens the mystery, about the cosmos and our attempts to discover more about it.'
Darragh McManus, *Irish Independent*

'An appreciative survey of the vast canvas on which physicists do their creative work – the entire observable universe, from the beginning of time to its end (assuming there is one).'
Graham Farmelo, *Guardian*

'It's rare that you get a book that connects Dante's *The Divine Comedy* to the Higgs boson and the geology of limestone cliffs, and this weaving together of two thousand years of intellectual thought is one of the many delights of this book. It's a hymn to scientific endeavour.'
Professor Mark Miodownik

THE CONSOLATIONS OF PHYSICS

WHY THE WONDERS OF THE UNIVERSE CAN MAKE YOU HAPPY

TIM RADFORD

First published in Great Britain in 2018 by Sceptre
An Imprint of Hodder & Stoughton
An Hachette UK company

This paperback edition published in 2019

1

Copyright © Tim Radford 2018

The right of Tim Radford to be identified as the Author of the Work has been asserted by him in accordance with the Copyright, Designs and Patents Act 1988.

All rights reserved. No part of this publication may be reproduced, stored in a retrieval system, or transmitted, in any form or by any means without the prior written permission of the publisher, nor be otherwise circulated in any form of binding or cover other than that in which it is published and without a similar condition being imposed on the subsequent purchaser.

A CIP catalogue record for this title is available from the British Library

Paperback ISBN 9781-473-65891-2

Typeset in Palatino by
Palimpsest Book Production Limited, Falkirk, Stirlingshire

Printed and bound in Great Britain by Clays Ltd, Elcograf S.p.A.

Hodder & Stoughton policy is to use papers that are natural, renewable and
sustainable forests.
to conform to th

Haringey Libraries	
NN	
Askews & Holts	12-Jun-2019
530	
	3400027082

For my family

Contents

Acknowledgements

Without the support of my family, the encouragement of my agent Will Francis, the rigour of my editor Juliet Brooke and the generosity of Drummond Moir, this book would not have happened at all. I cite the sources from which I quote directly, but for its substance and theme, I am in debt to the library of the Royal Astronomical Society, and beyond that, to every scientist I ever talked to, and every book I ever read. For its errors, I alone am to blame.

A Signal in the Void

At the end of November 2017, scientists and engineers based at NASA's Jet Propulsion Laboratory in Pasadena, California, fired up a motor that had not been used for thirty-seven years. They sent a command to a spacecraft to ignite a thruster to shift its position and alter the angle of its antenna. Since the spacecraft was 13 billion miles from planet Earth, the command, travelling at the speed of light, took more than nineteen hours to reach it. Then the engineers had to wait another nineteen hours and thirty-five minutes to find out whether their manoeuvre had worked. It had. Thrusters puffed bursts of gas from their nozzles for a hundredth of a second, to turn the spacecraft fractionally so that its antenna pointed in the desired direction. The command was a simple one, and its success had no immediate significance – it was a test and no more – but the achievement made the broadcast bulletins and news headlines.

That is because the spacecraft was *Voyager 1* – the first man-made object to leave the solar system. It is a twin sister ship to *Voyager 2*, the only observer to visit the outer planets

Jupiter, Saturn, Uranus and Neptune before it too began to head for the space between the sun and other stars. And in the course of this journey, *Voyager* had gradually become one of the icons of modern science. It is both an instrument and exemplar, not just of human curiosity, but of the joy – a delight that is intellectual, aesthetic and emotional – of exploring what is beyond the horizon, and beneath the surface of things.

All science, like all religion, like most history, like philosophy and probably all great art, addresses a set of universal, enduring questions: how did we get here? Why are things as they are? Where are we going? What does it all mean? Is there an ultimate purpose to our existence, or is what we can see around us just the result of a horrible accident, or a sublime one?

What science – and in this story, physics – does is take a little piece of one of those questions and, systematically and provisionally, deliver an answer. This answer on its own may help nobody and answer nothing. But physics goes on to another little question within that bigger question, and then another and then another, and sooner or later, the mosaic of little answers starts to deliver something of more substance: a pattern, a direction of travel, a model that seems to make sense. Actually, substance might not be the right word: we can never be sure that what we see is reality: we may be observing a mirage, or a reflection of reality, or just the silhouette of reality, as if a figure through an opaque glass door. Plato immortalised this possibility twenty-five centuries ago with his image of shadows in a cave. But even if the evidence tells us something incomplete, it still tells us something.

I am very fond of the metaphor of the world around us as a book, but a book only open at the present page. We have learned to read some of this book. It has a huge cast of characters in a confined landscape. We can infer a topography, sense a culture, interpret a plotline and thrill at hints that whatever is going on now might be just a twist, an interlude, an episode in some much bigger story. And from the adventures and conversations on the page open before us now, we have to deduce a number of things: are we reading history, or fiction? What happened at the beginning? How will it end? Does this story have a conclusion, or is it still being written? *Voyager* is an experiment in physics and cosmology that began forty years ago and has already helped us understand a great deal more about what we call home, the solar system, and this planet, the only place in the universe known to be hospitable to life. It is one of a procession of increasingly ambitious instruments designed to explore the order that underpins all creation. There are vast and separate projects to probe the nature of inner space – the structure of the atom itself – and the fabric of the void that we call outer space. This book is a love letter to physics, written by an amateur, and begins with *Voyager*, perhaps chiefly because it has been with us – and receding from us – for four decades, perhaps because its very existence is enduring testimony to 300 years of scientific Enlightenment, and perhaps also because it has already carried our dreams for longer and for a greater distance than any other scientific device.

I

Journey to the Stars

Consider it a getaway vehicle, the ultimate in escapism. It is on a journey to the stars, but no one would call it stream-lined or aerodynamic, no one would call it state of the art, no one would now describe its instrumentation as sophis-ticated, and no one now uses the language in which its computer code was written. Nothing about it is miniatur-ised. How could it be, when its memory is in the form of an eight-track digital tape recorder, spooling backwards and forwards from a fading power source that delivers energy measured in fractions of a watt? It is not in a welcoming place: the ambient temperature of interstellar space is not far above absolute zero. Against the backcloth of darkness and the speckle of distant stars far beyond the last outposts of the planetary system, it would appear motionless, because the only things against which you could measure its motion are themselves so far away. It is, however, one of the fastest things ever to have left Earth.

It is now receding from the sun at more than seventeen kilometres a second, more than 62,000 kilometres an hour. This is a speed that is difficult to imagine. Were such a spacecraft to fly low over Greater London – which, of course, at that speed it could not – or the New York conurbation, or the sprawl of Shanghai, it would make the traverse in little more than a second, and be gone long before a sonic boom announced the instant of arrival. Like Shakespeare's Puck in *A Midsummer Night's Dream*, it could put a girdle round Earth in forty minutes. It is so far the only surviving emissary from Earth, launched by dreamers with degrees in engineering, in mathematical physics and in astronomy, dreamers with patience, determination and a strong practical streak. It represents the last great ambition of the first years of the space age, the years 1957 to 1977 that saw the Soviet Union put the first object into space, the first dogs, the first man, the first team, the first woman, and launched the first missions to another planet; the years in which American ambition and enthusiasm put the first humans on the moon, and brought them back alive. For a few years, the Cold War seemed to take the form of a space race. Then it slid back into a war fought by proxy, with napalm and the defoliant Agent Orange in Southeast Asia, and governments on both sides began to lose interest in discovery for the sake of it.

But by 1977, the *Voyager* mission was a then-or-never project: it could not wait. It was dreamed up to take advantage of a planetary conjunction that happens only once every two centuries: all the outer planets would be on the same side of the sun, so one spacecraft could visit them all. The *Voyager* project survived political cutbacks: its authorisers budgeted for a four-year journey, but its makers quietly prepared for forty years. *Voyager* was launched with a letter from President Jimmy Carter and a song by Chuck Berry, and an onboard computing system with about one millionth of the computing capacity and capability of a modest modern smartphone, and with three separate computers instead of one, each of them vastly bigger than a smartphone, and duplicated in case of failure.

When it took off on a Titan rocket in 1977, *Voyager* was the climax of a dream. This dream was itself older than the space age, and some of those first dreamers of interplanetary exploration even lived to see it, and to see it succeed. It was a double mission: two spacecraft, *Voyager 2* first and then *Voyager 1* just sixteen days later, were launched with different flight plans, but between them they flew by Jupiter and Saturn and their moons, and then Uranus and Neptune. They carried with them an array of instruments: a small arsenal of detectors designed to capture light for photographs, to analyse electromagnetic radiation, to detect and identify any particles in the spacecraft's path. *Voyager*'s begetters prepared for almost every form of radiation, any kind of electric or magnetic field and any kind of cosmic particle hurled at the solar system from the distant stars, that they could think of.

Both spacecrafts used the massive gravitational field of Jupiter to accelerate itself to a speed that could take it out of the solar system altogether. This is commonplace now: the equations of physics deliver solutions that save on fuel and make epic voyages possible, so both spacecrafts entered into a trade-off with the biggest planet in the solar system. As each one looped around Jupiter it slowed down the planet's spin, but as part of the same bargain captured the equivalent force in the form of velocity. This was not a trafficking that should cause any concern. The energy of Jupiter's gravitational field accelerated each *Voyager* to dizzying speeds, and – since the laws of physics dictate that energy must be conserved – the drag of *Voyager* slowed Jupiter's spin by a rate that could be measured in centimetres over a period of a trillion years. Each *Voyager* was sent sailing into distant space by the Herculean heft of the Jovian gas giant: a discus launched by a Roman god. Both went on to Saturn, and then *Voyager 2* to Uranus and Neptune, *Voyager 1* out of the solar system altogether.

In February 1990, 6 billion kilometres from home and high above the plane of the ecliptic – the rotating, planet-bearing dusty disc of the solar system – *Voyager 1* was about-turned, to point its cameras for a last family portrait of the little worlds that orbit the sun. The sixty photographs are there in the world's album, and one of them inspired words that made my scalp prickle at the time I first read them, and still do so over two decades later: the astronomer Carl Sagan's hymn to the home planet, *Pale Blue Dot*, published in 1995.

> Look again at that dot. That's here. That's home. That's us. On it everyone you love, everyone you know, everyone you ever heard of, every human being who ever was, lived out their lives. The aggregate of our joy and suffering, thousands of confident religions, ideologies, and economic doctrines, every hunter and forager, every hero and coward, every creator and destroyer of civilization, every king and peasant, every young couple in love, every mother and father, hopeful child, inventor and explorer, every teacher of morals, every corrupt politician, every 'superstar,' every 'supreme leader,' every saint and sinner in the history of our species lived there – on a mote of dust suspended in a sunbeam.

Earth is, of course, barely visible at the distance from which the photographs were taken. A bit later in the same chapter, Sagan wrote:

> Our posturings, our imagined self-importance, the delusion that we have some privileged position in the universe, are challenged by this point of pale light. Our planet is a lonely speck in the great enveloping cosmic dark. In our

obscurity, in all this vastness, there is no hint that help will come from elsewhere to save us from ourselves.

When I feel more than usually desolate and wish to escape from the squalor and shame of so much human behaviour, I remember *Voyager* and realise that escape is impossible: we are stuck with the planet we have, and must make the best of it. In a world seemingly characterised by resentment, suspicion and hostility; by bigotry and condemnation; by barbed-wire borders and travel bans; by almost grotesque levels of inequality; and by greed, *Voyager* is a reminder that humans are also capable of selfless co-operation in pursuit of entirely unearthly satisfactions: if ever there was an instance of other-worldly pre-occupation, *Voyager* is it.

So some of us can still derive a measure of consolation – yes, a quantum of solace – by thinking of *Voyager 1* and *Voyager 2*. Each of them by now beyond the notional limits of the solar system, each of them heading into interstellar space, each alone and far from the other in the cold and the dark, most of their instruments now turned off, their tape machines turning back and forth to record and relay their 68kb maximum of data, both sending ever more feeble messages home, messages that can be picked up and read only by the world's largest antennae, both eventually to fall silent as their plutonium power sources become increasingly exhausted, and both bearing not just evidence of the civilisation that created them, but images, words and music, too. It represents an act of hope: that at some moment some alien civilisation, if any exists, will intercept one of the spacecraft, examine it, discover its gold-plated record bearing greetings in Earth's languages, the sounds of whales greeting each other, pictures of dolphins leaping, and the music of Beethoven and Chuck Berry. This notional alien discoverer may have difficulty interpreting the significance of the find: difficulty working out that the disc must be rotated at a certain speed while being scratched with a stylus which will then relay vibrations that only make sense when transmitted by air at atmospheric pressures that one might expect on a relatively small rocky planet covered mostly with water. But then, ironically, so would some people now on Earth.

Technology has moved on. Teenagers tend not to spin long-playing records. And programmers no longer use Fortran, or any of the languages that informed the first computing systems. Earth went one way, and the *Voyager* spacecraft another. *Voyager* is now an antiquarian emissary, a laughably overweight mission with clumsy systems that have this one virtue: that, despite an occasional hiccough, they have continued to work, no matter how improbably, for forty years in an environment that no living thing could survive for seconds. *Voyager* is in the grip of cold so fierce and deadly that we have no words for it, except 'nearing absolute zero'. It flies through a darkness for which we have only classical imagery: we call it Stygian and evoke the landscape of death and oblivion. And yet *Voyager* is alive, and going places, and it carries a message to the rest of the universe. And even if no one ever reads the message, or finds *Voyager*, there is this thought: never is a long time, and in an infinite universe, everything that can happen will happen, so ultimately someone *will* find *Voyager*, and read its message in a bottle. But by then, we'll be long gone, which I also find oddly comforting. Human ideas will endure for as long as humanity itself endures. The last thoughts will expire with the last thinking human. *Voyager* will endure as a palpable instance of such thoughts. It implicitly announces 'we were here'.

But why care? There are two reasons, maybe three, and they are rooted in physics. If physics is the search for the laws, the principles, the order that underlies and shapes all our reality – the air we breathe, the water we drink, the rocks we stand on, the solar energy and carbon dioxide that somehow become trees, and shrubs and grasses and of course, lunch – then an experiment such as *Voyager* is evidence that we know something about that order and want to know and understand more. We can do the sums, calculate the fuel, build a spacecraft, despatch it and watch it go. And keep on going. *Voyager* is both an assertion of confidence and an expression of hope. Those of us brought up to recite the Nicene Creed on Sundays would state, without thinking much about it, that among many other things we believed in the resurrection of the dead, and maybe some of us really did believe this to be true, but we could never have demonstrated such a truth.

Voyager confirms that a body subjected to a force continues in a state of uniform motion until another force is applied, and that those other forces can be predicted, and then checked against those predictions. In evolutionary terms, there is no obvious need for such knowledge: dolphins, eider ducks and dromedaries evolved on this planet and successfully occupied their ecological niches without founding any tradition of scholarship and without, as far as we know, formulating any questions about the purpose and meaning of life. But one mammalian species that evolved in and depends utterly upon the tiny biosphere of Earth, can nonetheless calculate velocities and predict flight paths that will despatch an emissary to the faraway stars. We talk of 'a thirst for knowledge' but such ready-made phrases won't suffice. Curiosity is not a thirst that can be slaked: it is a permanent condition. Every answer points to more and sometimes more profound questions. And knowledge, by itself, achieves nothing: we want more, we want something elusive, called understanding, or wisdom. We want both the big picture, and our place in it.

Voyager was also not just the fruit of one intellectual adventure but the launch of another: what *Voyager* saw, and sent back to Earth – an aurora over Jupiter's north pole, for instance, and tantalising hints of icy tectonic violence on the surface of Jupiter's moons Ganymede and Callisto, and even the discovery of a tenuous set of rings around the gas giant – immediately began to help us understand not just the planet on which we evolved, but profoundly more about the improbability of life on other planets, and about our extraordinary good fortune in being on this one, right now. Not only are humans alive, but we are alive to the colossal possibilities of life; not only can we ask questions, but – says the *Voyager* mission – we can answer them, or at least some of them, while at the same time finding new and ever more puzzling questions to address.

There is another special reward from *Voyager*. This is a confirmation that humans – collectively a species with a long and scary history of greed, selfishness, resentment and murder – are capable of dreaming of such a selfless venture and then co-operating to make it happen. We designed it, perfected it, built it, tested it, launched it and then nursed it from 1977 until now, but it's a fair guess that we didn't do it for the money. We did it because we believed in it. And if *Voyager* was, essentially, an American mission ultimately funded by the US government, it nevertheless advanced the dreams and ambitions of physicists and astronomers around the world.

Voyager is only one instance of the co-operative compulsion that seems to have been a feature of physics and astronomy from the very beginning, a collective and shared ambition to know more about the world and our place in it: there are others, and each of them testifies to the generosity of a big and often notionally mad idea.

Beneath the rocks that underlie the city of Geneva and the villages of France across the border is a tunnel, twenty-seven kilometres long, that is home to four vast experiments and a partnership of 10,000 scientists and engineers from all over the world: the European Organization for Nuclear Research, also known as CERN, home to what might prove ultimately an impossible attempt to resolve a philosophically ridiculous question: how did the universe begin? What strange phenomena took shape in the first violent microsecond of creation that made possible the world we now call 'mundane'? What determined the properties of matter?

In the US states of Washington and Louisiana an international partnership called LIGO, or Laser Interferometry Gravitational Wave Observatory, has built identical experiments – necessarily identical because to convince the researchers in the partnership, each instrument must confirm the other's data. While experimenters in Geneva want to interrogate the nature of matter, the substance of the universe, the LIGO research teams have begun to ask questions about the fabric of space itself. Their experiments exist only to detect distortions in space-time, at levels of sensitivity so extreme that across a length of four miles, instruments could register a shift of less than one thousandth of the diameter of the nucleus of an atom, and from this, identify an unimaginable event at unthinkable velocities a billion or more light years away, and therefore more than a billion years ago. All of these experiments have one key thing in common: there is no conceivable practical reward for the information sought.

No human being will be richer because *Voyager* has confirmed the existence of a heliopause, a region where the pressure of the wind from the sun – the shower of particles that identifies the solar system – is more or less matched by the pressure from the gas and dust from the distant stars. No government can possibly hope to profit, or advance its citizens' economic betterment in any demonstrable way, by the confirmation, at CERN in Geneva in 2012, of the subatomic particle that theoretical physicists proposed must have existed in the first trillionth of a second of time – when the universe was possibly about the size of a football, or perhaps a football pitch. Physicists 'needed' the Higgs boson to explain why matter has mass – their overarching model of why the universe is as it is required it to have existed in the first moment of creation. But theories demand confirmation at every point. So the identification of the Higgs particle was evidence that they had – so far – been working with the right model. This confirmation of its existence is limited: nobody could show you a Higgs boson, nobody could tell you what it would look like, supposing you could ever have seen it, and the evidence for it is still second-hand, because the CERN experiment in Geneva is capable, right now, of recreating only roughly the conditions that existed in the universe in the first trillionth of a second. Time, inside this first trillionth of a second, has already moved on, past the kingdom of the Higgs boson, and all that is left is the remnants of the remnants of the monstrous but mysterious object. The important thing is that the theorists knew what to look for, and then designed the experiments and built the instruments to find the answer, and then find it again and again, to their delight.

And the same intoxicated response occurred again in 2016, on another continent and in an entirely different international experiment, designed to answer a very different puzzle of theoretical physics. Although experimental and theoretical physicists are overjoyed by the detection of gravity waves – distortions in space-time predicted by the same model of why the universe is as it is – at the experiments in Louisiana and Washington, it is difficult to explain why. Their happiness was sparked by an event that lasted for twenty thousandths of a second, recorded by two instruments 3,000 kilometres apart within a hundredth of a second of each other. This was an event so small and so brief that no previous experiment had a hope of registering it. This same event told a story of two black holes, each with the mass of around thirty of our suns, that spun around each other at speeds rapidly approaching the speed of light, and then merged: and as they did these things, they distorted the fabric of space-time in a way that could have been predicted a hundred years ago from the equations of a physicist who started his career as a clerk in a Swiss patent office.

Notice that in each case, the physicists who claim a confirmatory result are not ultimately certain of their confirmations, and would like to see each find confirmed, again and again, and in different ways by different experiments. And notice, too, that even if they have a result, it isn't at all clear what that result really tells us about the universe we live in, and the matter from which we are fashioned. Physics works like that: theorists reason that if a hypothesis is correct, then a thus, or thus, would be an outcome, and if the outcome is confirmed by experiment or observation, then the hypothesis looks a trifle more sturdy. But it doesn't mean that the hypothesis is right, or the only explanation. And physicists can say such things, entirely unembarrassed by the realisation that they have spent billions of dollars in international partnerships and tens of thousands of career-years in pursuit of an answer that adds up to not just very little, but to a discovery barely distinguishable; a subatomic particle with a lifetime so fleeting it could hardly be said to exist at all, or a tremor in space-time so vanishingly small that it required the existence of two enormous detectors. That is, the gravity waves, those tiny tremors in space-time, were evidence of the same cosmic disturbance, and physicists could only be sure of this because the waves arrived, first at one detector, and then at the other, at exactly the interval you would expect if the event was real, and at the speed of light.

The latest reports from *Voyager* are even more nebulous. Looked at from one point of view, cosmic physics delivers nothing of commercial value and very little of philosophical certainty. Yet I find – and I know others do too – something enormously pleasing and exciting and, yes, comforting in these examples of the big adventure called physics, a way of trying to understand the universe. As J.B.S. Haldane said, in his 1927 collection of essays *Possible Worlds*:

> My suspicion is that the universe is not only queerer than we suppose, but queerer than we can suppose... I suspect that there are more things in heaven and earth than are dreamed of, in any philosophy.

II

The Carriers of Our Memory

Voyager is history. Its mission is all but over. By about 2020, or at most a year or three later, when its plutonium power-pack has decayed to the point where it can no longer deliver sufficient energy, that will be the end. The sensors may go on recording, and *Voyager* will keep on voyaging, but it takes power to transmit the data back to Earth. So for all practical purposes, *Voyager* will be dead to the world, or rather to this world. But it remains alive and important for this admirer, and quite a few others, for two reasons.

One is that it changed our understanding not just of the solar system but of ourselves and our home planet. It completed a survey of the planetary system, and in doing so it put Earth itself in context. After *Voyager*, we could begin to see the things we had in common with other planets, and understand all the better why Earth was different. *Voyager* was one of a succession of Soviet and US missions to our neighbours, and in time, we began to understand Earth as the Goldilocks planet: neither too warm nor too cool, but just right. It is the only one of the nine planets in which water can exist as vapour and ice, as well as liquid;

it has a hot, energetic heart and an ever-renewing surface, along with a magnetosphere shield to deflect or trap damaging radiation, and an atmosphere. Mars and Venus have atmospheres, but no liquid water. Jupiter has a magnetosphere, but one so powerful it represents a radiation hazard even to visiting spacecraft. Mars has a magnetosphere, but its internal dynamo switched off a billion years ago: it is effectively an inert planet. We now forget what we didn't know before *Voyager*, and didn't even know we didn't know, and the many things we thought we did know were wrong. So *Voyager* lives on: it played its part in reshaping humanity's image of the cosmos.

There is another palpable and immediate reason why it has achieved the nearest thing to immortality humans can imagine. It is one of the few man-made things that will survive not just the space age, but the Anthropocene, the era of humanity. It has escaped the gravitational field of the star that created the conditions for its own existence – we owe all our energy to the sun, and our homes, our history and ourselves to its hold on the dust and chemicals from which we are all made – and is beyond the reach of not just earthly harm, but solar injury too. It is falling towards other stars, but could easily miss them: *Voyager* may be travelling at over 62,000 kilometres an hour, but each of the stars of the galaxy is moving too, and any rendezvous is uncertain. Nor is *Voyager* likely to hit anything on the way: the universe is not merely vast, but vastly empty.

On the latest count the visible universe – and beyond the visible universe is yet more universe, so distant that the light from its stars has not yet reached us – is home to 2 trillion galaxies, each made up of maybe 100 billion stars, each star a thermonuclear fusion reactor, spreading its radiance across the universe. Each galaxy is also home to huge concentrations of mass that we cannot so easily see: cold and wispy nebulae of gas and dust, stars growing cold and dim, dead stars, stars so big their atomic structure collapses to densities so huge that a teaspoonful of ultra-compressed star-stuff could weigh as much as the *Titanic*. Beyond this there are black holes forever devouring the matter around them, and orbiting the stars are comets, asteroids, planets, planets with rings, and then the by-products of stellar fusion and stellar explosion: oceans of water, and strange accidental fabrications of organic chemistry such as alcohol and cyanide. And then there is the dust: cosmic Saharas of dust that may or may not coagulate into asteroids and planets. But this unimaginable mass of matter is still as nothing to the vastness of the universe that contains it. Combine all the visible matter in the universe and it still comes to a number of atomic particles that could be counted on a hand-held calculator of the kind given to schoolchildren in the 1970s that had a multiplication limit of 10^{99}. Spread all this unimaginable mass evenly across the emptiness of space and it would be invisible, and all but beyond detection.

Cosmologists have various calculations. One atom per cubic metre of space would be generous; others suggest one proton – immensely smaller than an atom – per four cubic metres. So even though *Voyager* has to find its way safely through the Oort cloud (the zone where the comets lie in wait), and survive possible encounters with entirely random, unidentified starless objects that might themselves be adrift between the stars, the probability of collision is effectively zero. *Voyager* is going somewhere, and there is no stopping it. Humanity, however, may be going nowhere. Humans may extinguish themselves within this century, through some folly or act of nihilism, or simply through slow destruction of the environment that gave them what they needed to flourish, to grow from a small group of foraging primates with sharp stones and the capacity to wonder, to a population of 9 billion or more later this century with intercontinental thermonuclear weaponry and a highly developed sense of personal greed.

Even if humans do not destroy themselves, they are doomed. Humans are an evolved species, and the lesson of palaeontology is that no species endures indefinitely: 2 million years, maybe, or 11 million years at most before *Homo sapiens* becomes a set of fossils, incarcerated with its own polymer products. The constructions of humanity will all crumble. Our lives were possible only because we live and are nourished by a star that burns hydrogen to make helium, like 90 percent of all the stars in the galaxy. Astronomers know enough about these creatures, called main-sequence stars, to understand their life cycles, and the star we depend on is at least halfway through its 10-billion-year lifespan.

That means that in less than 5 billion years, all the oceans will have evaporated, and in 5 billion years the sun itself, giver of life to all living things that we know of, will have puffed itself up into a colossal fireball, swollen to Red Giant status, to incinerate Mercury, Venus, Earth and perhaps even Mars. The inner solar system will become a dead zone. Not even the most enduring fossils will survive the solar incineration. Everything will eventually be dismantled into its constituent atoms and molecules, and continue to exist as the raw material that could ultimately be drawn into the orbit of another nascent star, and some new evolutionary cycle might commence, with some new intelligent civilisation. There is no way of knowing.

But through all this, *Voyager* could still travel on, cold, silent but entirely untouched, gleaming testimony from an entity that had perished utterly. Voyager tells stories of our history, our planet, our solar system and our universe. Its mute structure – a kind of saucer attached to a mug with some detectors kept awkwardly at arm's length – speaks volumes. Just as you could tell the history of the world in a hundred objects, you could tell a hundred stories from one object: *Voyager*. It is not just an artefact of science, it is its own *objet d'art*. It represents something astonishing, thrilling and aesthetic. A few of these hundred stories can be picked apart, one at a time, from the details of the spacecraft's conception, history, manufacture, launch, direction and payload. But *Voyager* is also a mission to the future that evokes memories of its own past (although the only people who will fully understand them as memories are alive now).

One of the sounds preserved on the famous twelve-inch gold-plated long-playing record aboard *Voyager* is that of the inside of a blacksmith's workshop. And one of the images is an artist's sketch of continental drift. Both tell us something profound about the world that dispatched the space probe.

Most people today have never seen inside a blacksmith's shop. In the mid-1960s, in rural Kent, when my children were tiny, we made a point of walking past the village blacksmith, to pause, admire the huge, heaving bellows, feel the heat of the forge, watch the smith pick up the red-hot metal with a pair of tongs, lay it on the anvil, and hammer it into shape. Then we'd hear the strange hiss as the farrier, still using the tongs, pressed the glowing horseshoe against the uncomplaining but sometimes impatient horse's upturned hoof – to an acrid smell of burning horn – and hammered in the nails. To be honest, it didn't happen very often, and our blacksmith was by then the only one for many miles, and he made most of his living out of wrought-iron garden furniture, door knockers and fireside hardware. Word would get out that on Saturday morning he might be shoeing a horse, and those who had never seen such a thing made a point of going to watch. It gave us the opportunity to observe the foundations of human civilisation: good or bad, it is not separable from the sound of the smithy. Neolithic people certainly had a civilisation: industry, agriculture, trade, religion and identity politics. They must also have had song but we cannot know anything much about the tunes that were sung or the words that were used because there was no writing to preserve the lyrics and the ancient songs faded into the ancient air. The story of the Neolithic people is reassembled and imperfectly interpreted from stones of the past.

Neolithic people must first have cultivated the ground in the Fertile Crescent by hand, clearing with a stone axe, tilling with a mattock made from deer's horn and harvesting with a sickle fashioned from flakes of flint. Once the smith had hammered a ploughshare, a team of oxen could be set to work. Something profound changed when imaginative humans found a way to wrest copper from the ground, and later smelt more of it from copper-rich minerals, and then mix it with tin to make bronze. Copper and tin are not companion metals. The former is more likely to be found amid bits of seabed basalt and sediment scraped up into a geological structure such as Cyprus, the eastern Mediterranean island from which copper derives its name. The latter is most often found in or near granite outcrops, great lumps of plutonic melt that rear up from much deeper, and push up great mountain ranges or other landforms, such as Cornwall in the UK.

The arrival of the Bronze Age is a story of the movement of peoples: the Bronze Age epics salute blacksmiths, and celebrate swords, armour and valuable dishes, but the same stories also tell us that Europe and the Levant at the time represented a vast free-trade zone, with free movement of peoples, especially skilled workers dealing in products of high net worth. You couldn't deliver trusty bronze swords or ploughshares without testing the proportions of each metal that made the alloy you wanted, the one in the highest demand, so locked up in the idea of the smithy is the concept of trial and error, or to put it another way, scientific research and development. You couldn't have a smithy without mines, smelting plants, sources of fuel, deliveries of ore, so the simple existence of bronze swords made to a preferred pattern, and bespoke armour, meant that the blacksmith had to be a specialist, who could use his earnings from the forge to finance his own shelter and secure his family's food supply. A smithy is a sign of the arrival of some sort of stable system of government, of collective food security with both pastoral and arable sources, and of intercity trade based on traffic by foot, by horse, by chariot and by wooden ship secured by copper nails, which did not corrode as iron does.

The arrival of the Iron Age was a profound technological advance, but an advance of a technology already available. Iron and then steel made better swords, and stronger ploughshares. There is a long and wonderful story locked in the sound of a hammer, beating on an anvil. It is the story of technology itself, from arrowhead to intercontinental ballistic missile, from compass needle to calibration coil, from bronze sword to cutting-edge science.

The local smith had once been one of the most important people in the community. But by 1965 he was just an occasional starring act in parish entertainment. Yet when *Voyager* was conceived, designed, built and then launched, in many parts of the world, and even in parts of Europe, people still rode and ploughed with horses, and employed them to pull milk floats and brewer's drays. These things too were sometimes by then more show business than economic necessity, but the occasional wistful milkman, the boastful brewer and the rag-and-bone man who collected old iron for recycling would clip-clop in every street, past households of people who had been born into a world in which measurements like 'horsepower' meant exactly that.

The calculations that took *Voyager* into space began with a dreamer called Konstantin Tsiolkovsky, a provincial schoolteacher who developed his rocket-propulsion equations on 10 May 1897, he recorded at the time, and in 1903 published the historic *The Exploration of Cosmic Space by Means of Reaction Devices*. The modern world had already dawned. Also in 1897, J.J. Thomson identified a subatomic particle called the electron, the German pharmaceutical firm Bayer developed a drug called aspirin, the automotive pioneer Ransom E. Olds founded a motor-vehicle company in Lansing, Michigan, Thomas Edison patented what would be a precursor of the movie projector. By the close of 1903, Orville and Wilbur Wright had successfully flown a heavier-than-air machine at Kitty Hawk in North Carolina. The twentieth century might be said to have arrived on time. But, of course, in those years even in the economically rich and technologically advanced countries, most of the world's work was still achieved by horsepower, and every community had a blacksmith. We are creatures of the past: we are always, to some extent, what we have been. Tomorrow's world takes shape against the backcloth of all our yesterdays. Visionaries predict what is about to happen; most of us don't see it coming. At the dawn of the space age, with the launch of *Sputnik 1* in 1957, the orbit of Yuri Gagarin in April 1961, and the launch of the race for the moon, nineteenth-century patterns of life survived.

Factories and offices needed employees to handle every detail of production and sale, and these workers travelled from their homes by bus, buying tickets from a drover or a bus conductor, or they queued at railway stations for tickets sold by clerks, to board trains managed and manned by an army of drivers, guards, signalmen and porters. And to maintain a punctual workforce, there was another, even more timetable-conscious army of tradesmen with cars and vans – bakers, milkmen, mechanics, journalists and printers – who delivered the great British breakfast of tea, toast and a daily newspaper. It was a world, too, of shipyards and shipwrights, docks and harbourmasters, seamen and stewards. The world's trade was loaded by hand and crane or derrick, stowed in a hold by dockworkers and delivered by freighter. Most people crossed the oceans aboard a passenger liner because that was the cheap way to do so: those select few who flew across the Atlantic still expected to depart from London, stop at Shannon in Ireland for a meal while the plane refuelled, stop again at Gander in Newfoundland for more fuel, before landing in New York. Steam still powered some of the trains, but increasingly, railways were driven by diesel or even electricity. It had become an automobile world. That was in part because the automobile industry still employed skilled workers who made cars on an assembly line, and paid their employees enough to enable many of them to buy their own cars. It was also in part because of simple economics: a horse was valuable because it could do the work of ten men, but it required the food of four of them as fuel. A tractor could do the work of a hundred men, and it required none of the food it helped produce. Instead it consumed something that gushed out of a hole in the ground: all you had to do was put it in a tanker and take it to a refinery. But you could still occasionally hear the sound of a smith's anvil, even if it was rare enough to make you stop and listen.

Politically, too, it was a strangely stable world: divided, but stable. In 1972, when NASA approved the *Voyager* mission to the outer planets, Spain and Portugal were dictatorships, Greece had been seized by a junta of military officers, Germany was divided into two separate and hostile nations, France and Italy had been torn by political riots, and Tsiolkovsky's Russia had ceased to exist. The Cold War was, of course, acceptable as a status quo because the alternative most obviously on offer was mutually assured destruction by thermonuclear war, delivered by intercontinental ballistic missiles developed for aggressive purposes.

The space age had emerged in 1957 in parallel, and in symbiosis with, plans for an armed conflict that nobody wanted but everybody expected, and the apparent but precarious stability of the Cold War stand-off. Most of Asia and Africa counted as the 'developing' world. Most of the oil-producing nations had, at the dawn of the space age, yet to discover the power with which mere geographical possession of a resource had endowed them, but by the time the last US astronaut stepped off the surface of the moon in 1972, the Organization of the Petroleum Exporting Countries had learned how to use this subterranean strength and plunged the developed world into what became known as the OPEC crisis. Crude oil fuelled everything, including political muscle. But humans went on measuring energy in terms of horsepower. There is something pleasingly symbolic in the inclusion on the record aboard *Voyager* of the noise of a smith's hammer beating on an anvil, a sound evocative enough to drive the rhythm of the opening of Wagner's *Das Rheingold*, a sound that places the machinery and structure of the *Voyager* mission in a long tradition of technological innovation, exploration and intellectual adventure.

The artist's diagram of continental drift included on *Voyager 1* is a simple one: it shows us the presumed shape of the ancient continental mass when Earth was 1.5 billion years old, in the form of the Pangaea supercontinent, and then the shape of the terrestrial world as it is now, at 4.5 billion years old, and then what it might look like a million years hence when, perhaps, an alien intelligence gets hold of the golden disc. But the sketch is also a statement about the discovery of our own planet. The choice of the phrase 'continental drift' is telling. It was the term applied much earlier in the twentieth century by geologists who argued that the continents must have been floating or dragging themselves across the earth. The phrase itself tells us how little we knew at the time about the surface of our planet.

But by that time scientists had begun to use ballistic missiles to launch technology to observe and monitor our planet. The next step was to use the same technology to visit other planets. But the books and journals published before the first missions tell us little about the other planets. Astronomers had used two centuries of systematic observation and measurement to deliver textbooks and reference works that defined the equatorial radius and mass of each planet and its distance from the sun, and the acceleration of gravity at the surface of Mars, Jupiter and so on, all of it in terms comparable to those on Earth. They also could tell you things like the sidereal period, which is the time taken for each planet to complete one orbit around the sun, which for Earth is one year – hang on to that bit, because it is important in this story – and the period of time taken to rotate on the planetary axis, which for Earth is one day. They also could confidently name the satellites they could see: none for Mercury and Venus, one for Earth, two for Mars, twelve for Jupiter, ten for Saturn, five for Uranus, two for Neptune and one for Pluto. We know now that the satellite count was incomplete. What we don't appreciate so much now is how little we really knew about the other planets, and how uncertain and confused we were about Earth itself. In 1957, at the dawn of the space age, geographers and geologists could not agree on any single, unifying explanation for the appearance of Earth's crust: for the pattern of volcanic eruptions and earthquakes, for the presence of marine sediments high in the Alps or Himalayas, nor for the strange fact that some mountains seemed to be getting higher, while other landscapes were subsiding.

Nor could humans be sure that Earth was the only home for life in the solar system. Arthur C. Clarke, now better known for his science fiction, was one of the post-war proselytisers for what, until 1957, seemed the impossible dream of spaceflight. He has a place in space history: he made the first published proposal for a geostationary communications satellite in the journal *Wireless World*, in 1945. His book *The Exploration of Space*, a layman's guide to rocketry and spaceflight, first published in 1951, speculated that if life existed on Venus 'it would be reasonable to anticipate much more advanced life-forms than on Earth, though possibly of a completely alien type, rather than creatures that might come from our own primeval past'. If intelligent beings existed, they would probably have quite a different history of scientific development, because they could never look out into space, so thick and so permanent were the Venusian clouds. (At the time, radio astronomers had still to develop the techniques that would answer some fundamental questions, such as the temperature on Venus – hot enough to melt lead – and the length of its day, which is 247 Earth days.) The surface of Mars, conversely, had always been visible: clear enough for at least one astronomer to claim he could detect canals on the Martian surface. By 1951, astronomers had more or less settled on a canal-free Mars, a waterless, cold, inhospitable place. But, said Clarke in 1951, there might be plants on Mars, getting oxygen from the soils, though 'the prospects for animal life appear somewhat gloomy'. Well into the 1970s, when the *Viking* mission placed a lander on Mars, space scientists were prepared for the possibility of some kind of life on Mars. They still are, but they don't now expect to find it on the surface.

If in 1951 we didn't know much about the other planets, we also knew surprisingly little about Earth. Shape, variation and the marine fossils on the tops of mountains were explained a little loosely. Continents rose and fell, were shaped and buckled by forces that began deep beneath, and then eroded by wind and rain, were submerged for long periods and built up again by sediments deposited in the ancient seas. The geological evidence testified to dramatic changes in climate and sea levels, and to seemingly inexplicable similarities in rock formations on different sides of an ocean, But geologists had yet to work out why these things happened, and in what order.

In a beautifully written and profoundly influential book, also published in 1951, Britain is dated to the end of the Palaeozoic period, 'embedded in the continent of Atlantis'. By the Cretaceous period, writes Jacquetta Hawkes in her best-selling history of our nation's real estate, *A Land*, the once narrow belt of that bygone ocean called Tethys 'now extended over much of Atlantis and was beginning to look like the Atlantic'. 'Britain still belonged to North America, to Atlantis, but already jaws of the oceans were waiting to close between Greenland and Europe and so to transfer the future allegiance of these islands to the European continent.' To help her readers understand nation-building as a geological process, she imagined a cine camera mounted on the moon long ago, its record now screened at tremendous speed. 'Towards the end of the last available reel the jaws of Scotland, the snouted face of Wales, the elegant Cornish toe, stumpy Kent and the bald head of Norfolk could be seen taking shape among the waves.' Continents rose and fell. They didn't move around. It is a given of geology that the lowest stratum is the oldest, and since the continents sat on the oceanic crust, the oceans must be the oldest features – creation's bedrock, so to speak.

By 1968, when Peter Ritchie Calder wrote a history of the world called *Man and the Cosmos*, geologists had entered another world, one in which – in the decade that followed the International Geophysical Year and the first launch of a satellite, and then a satellite carrying a dog, then two dogs, then a man, and then Soviet and US attempts to fly past the moon, and Mars, and Venus – the world picture had changed. Humanity had been able, for the first time, to look down on Earth from a distance, and within a year, someone would carry a cine camera to the moon. It had suddenly become clear and – though still not easy to understand – vivid: Earth moved. It shaped and reshaped itself; its skin crawled, and erupted, its crust was composed of a series of plates that banged against each other, dived under each other, and the impact of these stupefyingly slow but crushingly powerful forces created the landscape we see today, but will not see after a few million tomorrows. And locked within this one tiny visual artefact demonstrating continental drift posted aboard *Voyager* in 1977, one of many such diagrams, are another hundred stories of discovery and enlightenment. The history of science begins with the attempts to understand the planet we stand on, and its place in the heavens.

Just about the oldest working instrument in the scientific toolbox is geometry, or earth-measuring. In the third century BC, Eratosthenes of Alexandria worked out that Earth must be a sphere, and he did so from the observation that in

Alexandria at the summer solstice the sun cast a shadow with a measurable angle, but at Syene on the upper Nile far to the south, right smack on the Tropic of Cancer, it did not, because the sun was right overhead. So that meant Earth must be curved. All he had to do was work out the distance between the two cities, use the Euclidean equations for calculating a circumference and come up with a conjecture for the dimensions of the globe. There is an argument about how close he got to the real figure – it all depends on which version of the standard Greek measurement he was talking about when he quoted the answer in stadia. Eratosthenes is also credited with calculating the angle of Earth's axis of rotation, the tilt that delivers the four seasons. He wasn't the first person to reason that Earth and the other planets went around the sun: that honour is usually attributed to the somewhat older Aristarchus of Samos.

The little sketch of continental drift aboard *Voyager* contains within it a vast history of planetary understanding that extends back to Newton, to Copernicus, to Aristarchus of Samos and the story told about him in a text attributed to Archimedes of Syracuse, the first man credited with applying mathematics to the problems of physics. But instead of being just another digression, the Aristarchus connection is a reminder of why *Voyager* was launched in 1977. It could only have happened when it did because of a coincidence of orbital dynamics: something that happens every 176 years. All the planets in the solar system would be lined up on the same side of the sun, and in the nearest thing to a straight line, on 10 March 1982.

A bestselling book written at the time (by scientists who later wished they had not) even speculated that the combined gravitational tug from one direction of all the planets of the solar system might trigger various uncomfortable terrestrial phenomena, including a great earthquake on the San Andreas fault in California. No such horrors happened, at least none that could be confidently attributed to an assembly of other planets, all tugging in unison. But for a small team of visionaries at NASA, this planetary line-up presented a bicentennial opportunity. A spacecraft launched at the right moment could visit all the outer planets in turn, using the gravitational slingshot effect of each one to make it to the next. Miss that once-every-176 years bus, and the exploration of the outer planets would have to be piecemeal, and many times more expensive.

The coincidence of the planets was fortuitous: had this occurred ten years earlier, the mission could never have happened, because space-exploration technology was still in its suck-it-and-see phase. In the 1960s, Soviet and US engineers and scientists had certainly dispatched missions to Mars and Venus, but the initial success rate was not high: some spaceships never got there; others got there but failed to send back data. It was not until 1973 that *Pioneer 10* flew past Jupiter; it was not until 1974 that *Mariner 10* performed the first ever gravity-assist manoeuvre and purloined enough energy from Venus to change its trajectory and head for Mercury. *Voyager 1* and *2* (initially slated as *Pioneer 11* and *12*) could happen only because the engineers knew how to do it, and astronomers knew exactly when to do it.

All these visionaries and achievers had to do was to persuade the space bureaucracy that it was time to press the US government to put up the money and underwrite a Grand Tour of the outer planets. They had to do this while the US government was preoccupied with an energy crisis precipitated by the oil-producing nations of the developing world, with the political turmoil and human tragedy of the US wars in Vietnam, Cambodia and Laos, with the aftermath of the bitter civil rights campaign by African Americans that had climaxed with the assassination of Martin Luther King, and with the remorseless siege of the US presidency itself, by journalists, that was to end with the resignation of President Nixon and his replacement by Gerald Ford. The *Voyager* mission may have seemed like a never-again opportunity, but that was no guarantee that it would ever happen. Maybe the most amazing story is that it happened at all.

III

Adventures with the Time Machine

At a bad time in his life – in prison near Ravenna and awaiting execution – the Roman philosopher Anicius Boethius wrote a tract called *The Consolation of Philosophy*. It is a dialogue between the imprisoned and unhappy philosopher and the beautiful personification of Philosophy herself. In Book IV, she reassures him that she has swift and speedy wings

> With which to mount the lofty skies
> and when the mind has put them on
> the earth below it will despise:
> it mounts the air sublunary
> And far behind the clouds it leaves
> It passes through the sphere of fire
> From which the ether heat receives
> Until it rises to the stars.

Which sounds a bit like *Voyager*, except that the cosmology is very different and Boethius is in search of intellectual comfort rather than rigorous reasoning from first principles. This book is called *The Consolations of Physics*, a title that might be less impertinent if I knew a lot more about physics than I do, or if I'd ever demonstrated the kind of passion that might motivate someone to pass a physics exam at school.

The idea of truth as a consolation is a real one. We all face a death sentence, and we all want to believe in something that is demonstrably true. Religion in the shape of an institution, a set of dogmas and a body of doctrine, is not in this sense a help: it demands an act of faith, an acceptance that it must be true, even if, or especially because, empirical evidence is not available, and anyway, if you could see it was unarguably true, what would be the virtue of faith?

Physics – the kind of physics represented by examples in this book, such as *Voyager*, or the Large Hadron Collider at CERN, or the bewildering instruments devised to detect gravitational waves hurtling through the cosmos from the collision of two massive black holes a billion light years from Earth – represents an ambition for truth, a compulsion to explore reality, which sounds, if you say it quickly, so simple. 'Meanwhile, back in the real world,' people say by way of argument, but such arguments are very slippery. What is the world really made of? The Greeks proposed the idea of elements, and then one of them identified the last indivisible unit of matter as the atom. But even the simplest atom is mostly empty space, and divisible: when smashed in an accelerator an atom turns out to be made up of particles, and some of these particles don't seem to be very substantial, or very long-lived. And where did they come from in the first place? Reality is a strangely vague term. So perhaps physics is our attempt to fashion at least a simulacrum of reality, to achieve a glimmer of the real meaning of creation, if there is a real meaning we might be clever enough to comprehend. If so, a vehicle like *Voyager* or an instrument like the Large Hadron Collider would represent the definition of a sacramental oblation: an outward show of inward grace. That might be too fanciful. There probably isn't a moment of definitive understanding, of finality. The great scientific instruments represent more simply an earnest, palpable demonstration of the continuing search for an answer to why things are as they are. Newtonian physics got humans to the surface of the moon, it delivered a little lander to the surface of Saturn's moon Titan, it took a robot mission on a seven-year journey to an encounter with a comet no one had ever before seen, and it predicted the existence of Neptune.

But there is a catch. Newtonian physics is fundamentally wrong. It predicts the manifestations of reality, but does not explain them – gravity is a force that can be measured, but how does it work? How can there be 'action at a distance' between two heavenly bodies? And Newton's predictions only look accurate when applied to objects travelling at less than 14 percent of the speed of light. At greater speeds, very strange things can be seen to happen: time is different for those at higher velocities, and mass and shape seem to change at velocities nearer the speed of light. So by the beginning of the twentieth century, physicists began to look for new answers. Einsteinian physics, as manifested in the theories of Special and General Relativity, is now seen as a better representation of the forces of time, space and matter, but Einstein's version of the universe can't be the whole answer either, because there are things it cannot explain. It's a good working idea, so good you can sit down and commit money, brainpower, calculation, passion, invention and sheer determination to vast communal projects designed to check on the predictions that the theory might make.

One of these projects is the Large Hadron Collider at CERN. Just as *Voyager* was a mission driven by questions as old as Isaac Newton's celestial mechanics, CERN is an organisation that starts from the principles formulated by Einstein, his contemporaries and their successors. The $7 billion Large Hadron Collider is only the latest instrument in a procession of experiments at CERN designed to explore the ultimate nature of matter and, as a consequence, work out why the universe is as it is, and how it came to be. It does so by looking in detail at the nature of the hadron, a technical term with a precise meaning for physicists. A proton is one example of a hadron, and a proton is the hard bit in the nucleus of the atom. It is the bit with most of the mass. If you could work out why a proton has mass, and why it has the mass it does, you might learn something profound about matter itself, and the moment of creation. You can't smash a proton with a hammer, but you might be able to smash one with another proton, if you did so at enormous speeds. And with each collision you might be able to recreate the conditions that must have existed in the first microsecond of creation, before matter as we know it could take shape. So the Large Hadron Collider would become a kind of time machine, with which you could travel back nearer to the beginning of time and space and matter.

In 1997 I had the good fortune to meet someone with £1 billion worth of engineering contracts and a series of invitations to tender for contracts at CERN. In essence, he wanted to invite skilled engineering firms to make something that had never been built before: a series of superconducting magnets that would accelerate a hydrogen proton – the smallest lump of basic matter we know of – to 0.999999991 of the speed of light in a vacuum. To do this, the magnets had to be cooled to a temperature lower than the temperature of the space between the galaxies, and the twenty-seven-kilometre circular vacuum tube around which the protons would whizz would be the ultimate in earthly vacuums: it would be as tenuous as the atmosphere of the moon. And the proton traffic would be in two different directions (the clue is in the word 'Collider' in the instrument's name). To make two things as small as a proton collide requires engineering finesse taken to the extreme: CERN itself describes the challenge as like making two needles fired from ten kilometres apart meet head-on. Many of the engineers who built the working parts of the collider had to think in accuracies of a thousandth of a millimetre. Put at its simplest, engineering contractors had to make the emptiest space on Earth, and the coldest place in the universe, to stage the hottest collisions in the galaxy and they were expecting 800 million such collisions a second.

They also had to make the fastest thing on Earth. A physicist friend once worked out what 0.999999991 of the speed of light actually meant. If you stood on the International Space Station with an imaginary accelerator in one hand and a powerful torch in the other, and pointed both at our nearest star, Proxima Centauri, and pressed the button of each device at exactly the same instant, it would take four years for the first photons of light from the torch to make the journey all the way to Proxima Centauri. That's the distance, four light years. The protons from the imaginary portable accelerator would arrive just two and a half seconds later.

The speed of light in a vacuum is an absolute. Nothing can go faster, nothing can match it. The particles accelerated in the machine at CERN can approach ever closer to the speed of light, but never actually reach it. And they can reach the speed that they do only because the superconducting magnets that provide the acceleration are maintained at a temperature lower than that of intergalactic space, and fire their little pellets through a space as vacuous as the atmosphere of the moon. To evacuate the twenty-seven-kilometre tube that carries the hadrons, the engineers have to suck out the air, which would be rather like trying to extract all the air from within the great cathedral of Cologne, or Milan, or Strasbourg.

My cheerful engineering and particle physics procurement officer had a shopping list in his hand that stipulated the requirements for the superlative kit he needed. He wanted someone to bid for the contract to supply 8,000 superconducting magnets that would be more advanced than any superconducting magnets ever built, and since these magnets, some of them sixteen metres in length, would need to operate at a temperature of 1.9 degrees Kelvin, which is colder than minus 270 degrees Celsius, near to absolute zero, and certainly colder than interstellar space, he also needed a contractor to bid for the supply of 700,000 litres of liquid helium and eight 1,500m^3 containers in which to keep it, along with 12 million litres of liquid nitrogen to cool down 31,000 tons of high-precision metal instruments to near their operating temperatures. And then he wanted more mundane stuff: 50,000 tons of hot- and cold-rolled steel, 40,000 leak-proof pipe junctions, thousands of kilometres of superconducting cable filled with copper-titanium-niobium filaments, and all the additional extras, such as 6 million pairs of coil-clamping collars, and 30,000 copper wedges. All of it had to be of the highest standard, because once the Large Hadron Collider was up and running, on forty megawatts of power at its maximum, the momentum of the beam – just little puffs of hydrogen nuclei, accelerated to almost the speed of light in a vacuum – would acquire the relativistic energy equivalent to an intercity train travelling at 200 kilometres an hour. That is, if any one thing went wrong in this machine, everything could go very wrong indeed. As my engineering and particle physics informant said at the time, 'If you are not in the business of using superlatives on a regular basis, this is probably not for you.'

That, of course, is the second great attraction of advanced physics: everything involves something that has never been done before, with a precision and to a scale without precedent. This seems true for all the great ambitious science projects. The first attraction is the simple one of sheer selflessness. Whatever *Voyager* and the Large Hadron Collider are about, they are not about me, or us, or them. They are co-operative adventures in the pursuit of something bigger. The profit in the long term is real and profound and enduring, because the techniques and technologies pioneered in pursuit of the seemingly impossible sooner or later play into other, more general and more profitable advances. But the long term could be so long that no one would ever notice the connection, and so profound that no one would necessarily identify them with any particular advance in cosmic or particle physics. So the satisfaction is an emotional one: when we consider *Voyager* or the Large Hadron Collider or other big physics ventures, we are witness to the puny in pursuit of the apparently unachievable. Humans have joined together to do something that is both selfless and wonderful, in an attempt to discover a phenomenon predicted by theories most of us do not understand, backed by statistical reasoning we could not follow and technological invention we could not even begin to explain.

But there is a third satisfaction, and one in which we are all partners. *Voyager* and the Large Hadron Collider address questions we have all been asking since Bronze Age poets and visionaries composed the first Creation stories. The Book of Job in the Bible puts all this rather well: where,

God asks, was the sorry figure of Job when God laid the foundations of Earth, and then asks again:

> Whereupon are the foundations thereof fastened?
> or who laid the cornerstone
> thereof; When the morning stars sang together,
> and all the sons of God shouted for joy?

God has further questions for Job, covered in boils and sitting in ashes, his children slain and his flocks driven off:

> Hath the rain a father? or who hath begotten the
> drops of dew?
> Out of whose womb came the ice? and the hoary
> frost of heaven, who hath gendered it?
> The waters are hid as with a stone, and the face
> of the deep is frozen.
> Canst thou bind the sweet influences of Pleiades,
> or loose the bands of Orion?

Each of these questions demonstrates the presumed ignorance and helplessness of the human condition. The odd thing is that, in a practical way, physics has begun to answer them, starting from some very simple and easy-to-understand propositions. The Copernican principle, for instance, says that there is nothing very special about where we are or what we are made of. Copernicus himself did not actually make such a claim, and some scientists prefer to call it the Principle of Mediocrity, but it follows the logic first proposed in 1543 when Nicolaus Copernicus wondered if indeed, the sun and moon and all the stars revolved around a planet that represented the centre of the universe, and the focus of divine attention. The standard world-view in Christian theology at the time was that Earth was a little orb of imperfection occupied by fallible humans who play out a struggle for redemption, at the centre of a glorious amphitheatre of creation that becomes increasingly sublime and perfect as one ascends from Earth. The personification of Philosophy, in Boethius' *Consolation*, puts it beautifully:

> The sun into western waves descends,
> Where underground a hidden way he wends,
> Then to his rising in the east he comes;
> All things seek the place that best becomes.

But suppose, just for instance, Earth and the other planets instead revolved around the sun? If the sun was the centre of the universe, what would that mean?

The first thing it would mean – and this troubled Catholics, Reformation thinkers and people who thought Aristotle had resolved this nearly 2,000 years earlier – was that planet Earth was not the centre of the universe. And if so, therefore not the topographical centre of heavenly attention and, more troubling still, this could mean that the planets might be thought of as earth-like objects, in which case Earth might not be the unique focus of God's handiwork. It followed (although this took time and some practical help from Galileo, and from Isaac Newton, and then many others) that there might be nothing special about Earth.

In the centuries to come, it became obvious to astronomers that each of the distant stars was also a sun-like object, and each of them was at an immense distance. That is, not only was Earth just a physically insignificant stellar companion, like Mars and Venus and Jupiter and Saturn, but there could be a multiplication of such stellar systems, each perhaps with its own planetary companions. And this became obvious to these astronomers because they made a simple assumption: whatever physics said was true down here on Earth, must surely be true at a great distance. Collectively, individually, in piecemeal ways and systematically, astronomers and physicists began to test this assumption. Physics said that in a closed system, entropy increased with time; that everything went from order to disorder. This is the rule that says when you put warm butter in a cold fridge, the fridge becomes briefly warmer and the butter colder, until they are the same temperature. But the fridge is not quite a closed system: it is supplied by power to maintain a constant temperature. Nor is Earth a closed system: it is supplied by power from the sun to keep within a consistent temperature band.

The solar system is effectively a closed system. Although one day the sun will flare up and incinerate the inner planets, it too will die in the process, and the cinders of Mercury, Venus, Earth and Mars will then be the same temperature as Jupiter, Uranus and Saturn: frozen ghosts still wheeling slowly about the cold iron heart of what was once the sun. The solar system is not completely closed, though: starlight arrives, and comets from the distant fringe will occasionally make collisions that will manifest themselves as heat and light, and other stellar systems in the Milky Way galaxy could deliver, for a while, new light and life. The Milky Way could merge with Andromeda and then other galaxies too, so even the galaxy is not a closed system. But if the laws of thermodynamics are inexorable, then the Milky Way must one day begin to flicker, like a billion guttering candles. All sources of heat will eventually cool and dissipate. Everything will one day be at the same temperature. And since, effectively, heat is light, the galaxy, and then logically all the galaxies around it, will quietly grow dimmer and cooler, and then cold and dark. It will be just the same as oblivion, only with lumpy bits moving around.

And these frozen planets and dead stars will continue to move around because they are already moving, and one of Newton's laws of motion says that 'Every object persists in a state of rest or uniform motion in a straight line unless it is compelled to change that state by forces impressed on it.' This is of course the same law employed to say that if a mission took off from Earth at a particular hour on a particular day it would arrive at Jupiter exactly when its controllers intended because Newton's first law meant it was possible to predict exactly where Jupiter would be so many years or even many decades from now. And it was Newton's second law, which in its most economical form simply says that a body subjected to a constant force constantly accelerates, that delivered the conviction that a rocket fired from Earth with a payload of scientific instruments could outpace the downward acceleration of gravity and then sail blithely across the vast expanse of the solar system to a distant gas giant. These are all commonplace observations now, but until 1957, when *Sputnik 1* began its insouciant trilling as it circled the globe every ninety-two minutes, it wasn't obvious that it could be done.

The Copernican principle is an assumption: the laws of physics might be the same everywhere we can see, but what about that vaster arena beyond the horizon? What about earlier in the history of the universe? Or later? And perhaps Earth *is* special? Science has a way of making assumptions, and then questioning them, of testing the logical conclusions made from such assumptions. And in one respect, Earth might just be exceptional.

Right now and as far as we know – two huge caveats – it is the only place in the vast universe that is home to life, and not just life but one unique sentient form equipped with mathematics, curiosity and writing; a life form that possesses a consciousness that wants to know why the universe exists, why life exists, why we exist. In one sense this 'we' might indeed be living at a very special time and place in the universe. And that raises another huge philosophical question: did all this happen – and this is also the starting point for a number of religious traditions – just so we could make our appearance on the celestial stage? And if so, why? Were we in some sense intended? Could this vast and radiant universe have a purpose after all, and could that purpose be us, or could we at least be some way-station on the road to an even nobler project? Or if not that, some random experiment in the cosmic laboratory, masterminded by a higher intelligence? Interestingly, scientific research has a way of raising questions that circle back in the direction of religious thinking. The Bronze Age poet who composed the Book of Job put questions to God (who answered with questions of His own) and although science addresses and answers some of them in wonderful ways, it doesn't make the universe, or life, or us, any less mysterious and improbable. The Greeks had a word for that sort of thinking: *hubris*. But that doesn't mean it isn't true.

If we are the only living, sentient, communicating, wondering species in the universe, then we must be special. And if we aren't the only intelligent species in the universe, then where are all the others? This is another great question, put by the great physicist Enrico Fermi: where is everybody? We listen, but we hear no word from the creatures on other planets in this galaxy, or beyond. We despatch messages – we have been doing so in the form of television channel signals now for more than six decades and the first episodes of *Coronation Street* and the *Ed Sullivan Show* left the planet broadcast as microwave radiation and have already reached star systems that are home to planetary systems, some of which are known to be Earth-like – but has anyone responded with their own version of the romantic comedy, or the variety show?

As I write these words, yet another scientist has addressed the same question: in the *International Journal of Astrobiology*, Daniel Whitmire, a mathematician at the University of Arkansas in Fayetteville, wonders if what he prefers to call the Principle of Mediocrity – the so-called Copernican principle that says there is nothing special about us – holds true? What if technological species pop up all the time, in all appropriate locations, everywhere in the universe, and when they do evolve, they change the conditions of their own planet so drastically that they obliterate themselves and perhaps all life on that planet within a century or two, leaving little or no trace of their own existence, beyond the fossilised rubble they created and then destroyed? The notion that humans could in some way destroy the living conditions that encouraged the evolution of a contemplative and curious bipedal mammal with a gift for malevolence, metalwork and ballistic missiles has been around for a long time. St John of Patmos sketched an outline of apocalypse in the Book of Revelation, and twenty centuries later, Britain's Astronomer Royal, Martin Rees, composed a much more detailed and practical manual for the elimination of human life in his book *Our Final Century* (2003). He intended his book as a warning, and some of the dangers he cites – among them the quick fix of global thermonuclear war and the less intentional slow death by catastrophic climate change triggered by human profligacy – have been obvious for decades, and the global political climate at the time of writing makes some of these catastrophic conclusions seem more than usually probable.

If humans are nothing special, then perhaps this endpoint is inevitable: the species plays with wonderful ideas, creates wonderful but destructive things, and then all too quickly eliminates itself, and perhaps all life on its home planet. If it doesn't, then any intelligent technological species that does re-emerge from the smoking rubble will, millions of years later, repeat the process, and once again, no one would know that these earlier species had ever existed. Although if we are typical of intelligent life in the universe, then in a local universe packed with hundreds of billions of galaxies all containing hundreds of billions of stars, perhaps many of these alien and self-destroying civilisations will also have launched emissaries such as *Voyager*. Even if all sentient and outreaching life, everywhere in the universe, typically destroys itself, the silent universe might still contain the whispers of what might have been, in the shape of mute and enigmatic spacecraft, proceeding at fixed speeds across the darkening void.

I'm with Boethius on this one: there is consolation in philosophy, even if it is limited to just thinking about thinking, about formulating new versions of those great questions put to Job in the land of Uz. Maybe we are in some sense special. But that wouldn't stop astronomers, physicists, cosmologists or even mathematicians from backing the Copernican principle. That is because it delivers. It provides a sense of order, of predictability.

This brings us to another successful prediction to be made from this very simple cocktail of observations involving the Copernican principle, Newton's laws of motion, the telescope and the laws of thermodynamics: together they seem to say, if the universe is going to die, effectively of old age, then surely there must have been a time when it was young, or newborn? The book metaphor supplies the answer. Of course, of course, a book has either an author or an editor, so it has a beginning, and the only puzzle is: what kind of beginning?

Long before the space age, some cosmologists proposed that time, space, gravitation, light and energy all emerged from a 'cosmic egg'. The begetter of the egg metaphor, a Belgian priest and astronomer called Georges Lemaître, compared the evolution of the universe to a firework display that had just ended: 'Standing on a well-chilled cinder, we see the slow fading of the suns, and we try to recall the vanished brilliance of the origin of the worlds.'

But having discussed at length the idea of an expanding universe that began from a singularity, an infinitely dense, hot point at a distant time, the same book that introduced me to Lemaître also devoted an equally long chapter to the alternative theory promoted by the British astronomer Fred Hoyle and others, that the universe could be infinite and eternal, and sustained in a steady state by continuous but unobserved creation of matter. This steady-state theory would explain why the universe always seemed to look on a large scale the same in every direction. That is, in every direction there are galaxies, and voids between the galaxies, in roughly the same pattern and number. This thesis of continuous creation posed no intellectual problem. If you could create a whole universe from nothing and nowhere all at once, you could just as easily do it a little bit at a time. Either God could do it – the default explanation for most people for most of the last 2,000 years – or the laws of physics could explain it, or God could have employed physical laws to do His work for Him. The point is that in 1961, four years after the launch of the journey into space, nobody could possibly have known which explanation might be right. Theories make sense when backed by evidence, and until the mid-1960s, nobody knew what evidence to look for. Many physicists did not regard cosmology as science. 'There is speculation, and then there is wild speculation, and then there is cosmology,' I heard a physicist once say. It was not until 1965 that a set of radio astronomers proposed in the *Astrophysical Journal* that if – it was still a big if – the universe began in a big bang, the 'echoes' of that big bang would be everywhere. In a closed universe that was expanding at a colossal speed, the first fierce and unimaginably hot blast of radiation would still be rattling around in space, getting fainter and weaker and cooler and would now be at or no more than about 3 degrees Kelvin (i.e. 3 degrees Celsius above absolute zero). That is, the temperature of empty space would serve as a clock: what began as a hot, dense fireball would cool as the universe expanded: the greater the space between the stars and galaxies, the cooler the temperature of that space. Once again, the Goldilocks fairy tale becomes helpful: the temperature of the porridge in the bowl can tell us how long ago it was made. In the same issue of the *Astrophysical Journal*, a second set of communications scientists working for Bell Laboratories announced that they had detected a ubiquitous radio signal, indicating a temperature of 3 degrees Kelvin in the space between the stars. They had observed it everywhere they had looked: they had initially been puzzled as to the source, and had even (they didn't use these words in the scientific paper) wondered if the cause might be radiation from pigeon droppings that had dried fast on the bowl of their radiotelescope's dish receiver.

Now, fifty years on, that match of prediction and confirmation sounds like clinching evidence. As usual, it took maybe a decade for the idea to take hold, but relying on the Copernican principle that the laws of thermodynamics and motion and acceleration and gravitation would be the same in the past as they are now, and the same everywhere, other physicists started to test their reasoning and had begun to identify a timetable for the beginning of the universe. By the time *Voyager* was launched in 1977 most physicists were fairly convinced that the universe had had a beginning, somewhere between 10 billion and 20 billion years previously. That same year Steven Weinberg wrote a bestseller called *The First Three Minutes: A Modern View of the Origin of the Universe*. In the final pages, he also added that it was hard, living here on Earth, to realise that:

> all this is just a tiny part of an overwhelmingly hostile universe. It is even harder to realize that this present universe has evolved from an unspeakably unfamiliar early condition, and faces a future extinction of endless cold or intolerable heat. The more the universe seems comprehensible, the more it also seems pointless.

We didn't know it at the time, but *Voyager* was our emissary into the rest of the overwhelmingly hostile universe, and into the region where interstellar space itself has a temperature of about minus 270 degrees Celsius. When the spacecraft set off we had hardly begun to understand the universe we are living in, we still had no explanation for why it looked the same – on the largest scale – in every direction, and the guess for the age of the universe had a 10-billion-year margin of error. We now have an explanation, supported by decades of cumulative astronomical observation, for the homogeneity of the universe: it is a consequence of an event that is truly unimaginable, and that happened so far back inside the first second of creation that the numbers don't make any sense at all to ordinary humans. The best estimate for the age of all time, space, matter and energy in what we call the observable universe is about 13.8 billion years. And most of it is shining blackly at a temperature that defines the age of everything that is, or will be. And, since most of the universe is empty space, there's no reason why *Voyager* should not travel on for another 13.8 billion years, or longer still. There is, however, a good chance that even with its forty-year head start, it will be overtaken by something else dispatched from Earth, and likely to outlive Earth, the solar system and even the Milky Way. For me, *Voyager* remains special: it has travelled the furthest, and carries technology that tells, directly or implicitly, the whole history of physics, of wonder about the universe. It has a unique place in history, and compared to twenty-first-century spacecraft, it is big and reassuringly ugly and awkward, a child of its time. It remains a symbol of the great intellectual adventure of physics. But, of course, it is not the end of the story.

IV

Contemplation in Free Fall

Let us return to the idea of escape, a recurring theme in fiction. In science fiction, it often takes the natural form of the spaceship not as a voyaging Argo, a Magellanic exploration of the unknown worlds, but as a lifeboat: an ark carrying survivors from a planet about to be destroyed to some possible better place to start again. Our imagination says we can do it; our technology says we know how to do it: twelve astronauts have walked on the moon and there are people alive today who hope – and some even expect – to see a human settlement on Mars. Humans may occupy this Martian settlement only briefly, but so what? We have begun to reach out to the Great Beyond, and why should this reaching not go on, and go further? *Star Trek, Red Dwarf, Star Wars,* a 1950s BBC radio favourite called *Journey Into Space* and other cinematic and broadcast fantasies may indeed be fantasies, parables that tell us more about life on Earth than about life around the distant stars, but they are evidence of an impulse to explore. And we have the propulsion. All we need is to work out how to go further, and then further still, and then one day, all the way. Once people said that we could never fly: look at us now!

Any spacecraft that supports life must carry its own atmosphere, its own water, its own food. It must have motive power to accelerate and decelerate, and to change direction. It must have living space for humans to stay active, because (compared to an astronaut) even the most indolent of us must exert muscular pressure just to remain standing. We bear the crushing burden of the atmosphere in which we evolved, we resist or work against the downward acceleration of gravity at ten metres per second per second, and we do these things in a global air-conditioning system that is available on demand – and we make this demand every four seconds, as we breathe the happy mix of nitrogen, oxygen and a few trace gases that keep us just so, unthinkingly alive.

Each perambulating human is a walking tower block for invisible forms of life: 100 trillion or so microbes in thousands of species, with whom we have made some kind of evolutionary bargain: they live on us, but we also live because of them. Carbon-based animal life forms breathe in oxygen and exhale carbon dioxide which is then recycled and returned as oxygen by a sublimely reliable invention called photosynthesis, and this process is managed by a spectacular variety of plant and algal life that has never in the past failed to deliver, and is unlikely to fail as long as there is sunlight and water, nitrogen and a few other elements, not because plants can't fail, but because evolution has ensured such a variety of plants and algae that any individual or local failure makes no difference. The same plants also manage the hydrological cycle from which all life forms benefit. The *Apollo* astronauts who went to the moon did so with a mechanical system to recycle the air and a series of packed lunches. The astronauts on the International Space Station have grown salad and other vegetables, but largely as a scientific experiment, a demonstration that it can be done, and as a way of learning how best to do it, and for provender are supplied at intervals with fresh food and water from Earth, technically no more distant than London and Edinburgh, or Washington and Baltimore.

But the astronauts who go to Mars are going to have to take their own miniaturised version of Earth with them, in the form of a greenhouse forest of plants that will absorb the foul air and excrement of the animals on board, and supply nourishment in the form of harvestable food. And in this sealed environment, the earthbound rhythms of evolution and adaptation will continue: who knows what the microbial flora and fauna of a sealed spaceship will be like, after a few years, or a few decades?

The medical profession already knows what will happen to the space voyagers: crew, or passengers, pioneers or refugees, each human aboard a spacecraft will start inexorably to lose bone density, to find that body fluids have a tendency to hang around the heart rather than flush through the limbs, will start to demonstrate – however carefully the psychological screening has been – a sense of stir craziness, of exasperation at such confinement. The atmosphere and the ionosphere and the magnetosphere of planet Earth protects its denizens from lethal radiation: on any space voyage, radiation will be an enduring hazard that becomes more dangerous with increased exposure. Shielding will add to the mass of the spacecraft. The Martian explorers will require their own living and moving space, somewhere for occasional privacy, somewhere to store the tools and equipment they might need on Mars, recreational resources in the form of music, books and movies, somewhere to change clothes, to wash and sleep. Any spaceship that carries them will have to be a combination of warehouse, gymnasium, workspace, living space, garden and greenhouse, cafeteria, bathhouse and nuclear-fallout shelter. Designers must also make space for emergency supplies of air and water, and fuel.

And the spacecraft will have to get off Earth in the first place. The economics of a space launch are unforgiving: just to get into space at all, that is, into low Earth orbit, a launcher must generate a final speed of five miles (eight kilometres) a second. This is not just faster than a rifle bullet, it is nine times faster than a rifle bullet. NASA used to claim that it cost $10,000 just to put a pound of payload – think of a box of teabags or a small pack of sugar – into orbit. It also said that to put the space shuttle into orbit required a main rocket engine capable of generating power equivalent to the wattage produced by thirteen Hoover dams, and that thrust needed to be backed up by two solid rocket boosters which each at lift-off burned the fuel, every second, that would power 2 million saloon cars. At the time of writing, chemical rockets are the only thing that could put a mission into orbit, and then generate the extra speed required to put the spacecraft on an orbit around the sun.

Once largely free of the planet's gravitational tug, thrust becomes less expensive. Indeed, it might be achieved at almost no cost at all. It would be notionally possible to harness the power of solar radiation: to think of the spaceship as a sailing ship and hoist a vast, ultra-thin sail and ride along on the propellant power of starlight. Fuel is a problem for journeys through limitless space, and the spacecraft must carry all it needs, because there are no refuelling stations on the highway to the heavens. Light, however, is everywhere. In 1989, in yet another of those meetings with remarkable men who most people have never heard of, I encountered Louis Friedman, director and co-founder (with the astronomer Carl Sagan) of the Planetary Society, who while at NASA had done the initial theoretical calculations for a solar sailcraft. Sunlight itself is a force. Down here on Earth that deliciously powerful sensation on a sunny Mediterranean day is more warmth than impact. Up in the vast liberation of free fall, the impact of sunlight is measurable. Photons slam onto a surface, and bounce off. Make that surface a reflecting mirror and each collision follows Newtonian mechanical laws. Make the mirror into a fine sail big enough and thin enough, make the ship that hoists this sail light enough, and the light from the sun begins to act as a breeze. Under the pressure of radiation, the sail would begin to move, taking its ship with it. It would not just begin to move, it would begin to accelerate, because sunlight is a constant force, and a body subjected to a constant force constantly accelerates. Could you configure the sail area, and its tilt, and make the payload small enough, so that your spacecraft could gain an acceleration of one millimetre per second per second? Then in just twenty-four hours, this free-fall windjammer would be scudding along at a hundred metres per second and would have logged a journey of about 4,700 miles or about 7,500 kilometres, and the acceleration would continue.

Any spacecraft is a discovery machine: before it can go anywhere to discover anything, its designers have to solve the problem of how to power it. One notional design for a sun-driven spacecraft, drawn up by enthusiasts at NASA's Jet Propulsion Laboratory in Pasadena,

California, could create the speed needed to leave the solar system altogether, to report back from a world far beyond the empire of the sun. It could cover forty astronomical units a year. (An astronomical unit is the distance from Earth to the sun, a distance that, even at the speed of light, takes eight minutes to cross.) At such a speed, the space clipper would be travelling at more than a hundred kilometres a second – far faster than *Voyager*. It could cover the emptiness between Earth and our nearest star, Proxima Centauri, in about 6,600 years. Yet, there are problems. The energy delivered by radiation falls away as the square of the distance. By the time our spacecraft got to the asteroid belt, its rate of acceleration would have begun to flag. By the time it got to Jupiter, the solar energy delivered to its photovoltaic cells would also have begun to weaken, and by the time it got to Saturn, the electrical supply needed to keep the passengers warm, and cook the potatoes grown in the clipper ship's greenhouse, would have failed. The sunlight needed to drive photosynthesis would be negligible, and although plants would continue to harness the energy pulsing from a light-emitting diode in this faraway, fleeing greenhouse, the energy to make the current to light up the soar-away vegetable patch – think of it as the last word in heavenly plots – would have to come from some other source, which would have to have been aboard at launch, because on such a journey, there will be no ports of call, and not enough fuel to afford the deceleration necessary for a pit-stop pause, even if it were possible. There is wonderful imagery in Milton's *Comus* that seems to anticipate the true spaceship, riding away on a sunbeam:

> To the Ocean now I fly,
> And those happy climes that ly
> Where day never shuts his eye,
> Up in the broad fields of the sky

But by now it should be clear that nobody will be sucking 'the liquid ayr, all among the gardens fair'. The first solar sailing ship was hastily sketched – but never built – as a potential candidate for a rendezvous with Halley's Comet on its last visit to planet Earth in 1986. The spacecraft could have hoisted sail, and in nautical terms 'come alongside' the comet as it entered the inner solar system and curved around the sun: it could even have despatched a robot boarding party. But it was never built because there was no way it could have been launched, except aboard the space shuttle, which at the time of the great plan for a rendezvous, had itself not been completed. The solar sailor remains, for the moment, a beautiful dream, and there are no plans to send one scudding across the cosmos to the nearest stars. The dream of space exploration by radiation-drive has metamorphosed into something even more ambitious, but also harder to imagine.

In 2016, a Silicon Valley billionaire called Yuri Milner telephoned me to tell me about a better way to get to the stars. I was back on the science desk at the *Guardian* during a former colleague's absence. Milner's public relations agency had supplied him with a list of science journalists to brief, and he had chosen to do the briefing himself. He had a great story to tell. He wanted to put new life into what, by 2016, was an old idea of sailing to the nearest stars. Yes, he wished to see spaceships, with sails. And they'd be driven by the power of light. Only this time, they'd get to the stars in the Centauri group in about twenty or so years rather than many thousands of years.

His plan was simple: make the spacecraft very small, and the sails very thin and light, and then hit them with a powerful laser, because sunlight would be too slow to deliver the acceleration for a high-speed, quick-gratification starshot. He had a gift for making such propositions seem reasonable, but then for some of us they already sounded reasonable, if only because the idea of laser-assisted motive power had been around already for at least three decades, and had been proposed (entirely as a thought experiment) by the NASA teams that worked on the idea of solar sailing in the first place. But, Milner argued, once again perfectly reasonably, things that were happening at the time we spoke would have seemed wildly implausible two decades beforehand. So now would be the moment to start figuring out what we could really do in fifteen or twenty years. Moore's law – the one that says the computing power of a silicon chip could double in less than two years, every two years – had already enabled the compression of telephone, television set, tape recorder, video recorder, internet search engine, satellite navigation, pocket calculator, global atlas and other handy little tools all into one hand-held device. Orbiting observation satellites that once would have been the size of *Voyager* or *Sputnik* had been reduced to CubeSats – packed into a cube measuring 10 centimetres.

Space missions can fail: they can fail because the rocket falters and falls back to Earth, or because it explodes at high altitude; they can fail in orbit because the last stage fails, or because the technology aboard goes wrong, or the software malfunctions, or because of some random accident. And that's the end of that: the scientists and engineers behind the project write off the dream, and start again. Once upon a time the loss of a spacecraft represented the failure of a billion-dollar gamble: now you could pack half a dozen identical tiny spacecraft around some other larger instrument and not just save money on launch, but be sure that at least some would work. Perhaps by 2030 or 2035, with a bit more help from nanotechnology – the art of making machines on a scale of billionths of a metre – you could get every sensor now aboard *Voyager*, and maybe a battery and transmitter as well, packed into something that weighed only a gram or so. The same combination of nanotechnology and the new science of metamaterials – the construction of fabrics using just one layer of atoms at a time – could take care of the sails.

Lasers, too, have come a long way since the days when an urbane screen villain dreamed of using one to bisect James Bond. Their use is cheaper, the applications more flexible and the generation more powerful. And engineers now know how to lock lasers together so that they operate as a single device, delivering many times the power of each one. So, said Milner to me over the phone, it should be possible to hurl a flurry of tiny sailcraft into orbit, point a laser beam at them in the right orientation and right direction, and accelerate them with a high-powered battery of purpose-built lasers. Think of just one of these sailcraft: an ultra-lightweight sail the size of a kite with a starchip rather than a starship attached, weighing no more than a sheet of paper, out there in the high freedom of empty space, suddenly being blasted in the direction of Alpha or Proxima Centauri by a laser packing the power of 100 billion watts. It would indeed accelerate, for the very short length of time a laser could track it and drive it – just minutes, but in that time and if things worked out, the little voyager could be powered up to nearly a quarter of light speed at 37,000 miles per second or nearly 60,000 kilometres per second. It could cross the 40-trillion-kilometre divide to the Centauri group in fewer than twenty years, and start sending back information which would reach Earth within the lifetime of at least some of the scientists who designed and launched the project. For the first time, an earthbound species would have left its own star system and begun to explore an alien world: we think we know what a planetary system around another star might be like, but would it be like the one on which we evolved? Or would it be in some way unimaginably different? Could such a distant planetary system be home to life? And – since it would be hardly more expensive to make and launch twenty or 200 such spacecraft as to make one of them, and once you had such an amazing laser array *in situ* you might as well use it more than once – you could go on and on firing your little missions to the distant stars. If one failed, there was always another that might deliver, not much more than one human generation onwards. The word science derives from the Latin *scientia,* for knowledge. Art comes to us from the Latin *ars/artem,* which describes making something, fitting something together. Art and science are partners: an artificer made the first telescope as a fairground toy. Galileo turned it into an instrument to see the moons of Jupiter, and from then on, astronomy demanded ever better instruments. Astronomers and space scientists have expanded our horizons – which means they have simply shifted them to ever greater distances. If we want to see beyond these new horizons, we must devise new instruments.

It would not be simple. Milner and his partners and their scientific advisers had thought of at least twenty challenging problems to be addressed, among them the interesting puzzle of how much and what information such a mission could record as it sped through the Centauri group at nearly one quarter of the speed of light, and then how to organise the data and imagery and get it all back to Earth, all from a mothership weighing a few grams. The radio signals that *Voyager* transmitted back to Earth from Neptune, picked up by the world's largest and most sensitive antennae, were so feeble that even an electronic digital watch of the time would have had 20 billion times the power level, according to the Jet Propulsion Laboratory in Pasadena.

In his 1991 book *Far Encounter*, one *Voyager* historian, Eric Burgess, imagined the challenge of funnelling a stream of digitised data back to Earth, for example from just one photograph compressed to 2 million bits of data. By the time the last of the data from that one image left *Voyager*, the first bit of data would have travelled 17 million miles. That is, each image would represent a speeding arrow of information 17 million miles long.

Now try to think of the challenge of firing an even greater volume of information back to Earth from a star system – even the nearest stars – so distant that only the sun would be visible, rather than the third planet from the sun. Try to imagine what information such a mission could collect, as it whizzed past its target at such a speed. It doesn't matter right now if we can't imagine any such reward: we don't have to. All space missions have told us something useful or wonderful, even some of those that on the face of it might seem to have failed. The potential rewards, entirely in terms of intellectual excitement and discovery, are so great that Milner announced that he was prepared to put $100 million on the table to kickstart the research and experiment that might one day take a spacecraft to a distant star, and then beyond that to even more distant stars.

A theme emerges from all this: the theme of light. We live in a radiant universe. Ultimately, sight is our prime sense, the alpha and omega of awareness. Sound and touch, taste and smell – the music of the ocean waves, the mewing of seagulls, the buckshot sting of sand blowing across a hot beach, the scene of hot pines, the taste of olives at a seaside bistro – are all confined phenomena. They have evolved because we are lucky enough live on a warm, moist planet with an atmosphere renewed on a daily basis for us by photosynthetic plants, all competing for their place in the sun, every one of them a tiny solar-powered factory intent on fabricating chemicals that will fuel its competitive drive. These chemicals make it more attractive to animal predators that will help it reproduce, and more repellent to those that constitute a long-term threat. This can happen only because the animals too – bees, ants, butterflies, bullfinches, bison and barristers – have evolved the capacity to appreciate, gratify or protect the other senses.

The world is driven by light, and the senses of hearing, taste, touch and smell so important to us are mediated by light in the form of electromagnetic currents transmitted through nerve cells to the brain. All our senses but one are earthbound and atmosphere-dependent. In space, nobody *can* hear you scream. About a hundred vertical kilometres from where you sit, the only information-delivery system is light. We live in a vast and expanding universe, and the only way we know what we know about it is transmitted by waves of light: radiowaves, microwaves, waves of light that fit within the violet-to-red colour spectrum available to humans, and then beyond the ultraviolet to all the high-energy packets of light, the X-rays and gamma rays normally screened out by the protective atmosphere and magnetosphere of planet Earth.

And, ultimately, light may be all that matters. We seem to have always known this: the Book of Genesis opens with 'Let there be light' – one of the first moments of Creation.

Genesis, though, doesn't quite square with the evidence delivered by cosmic physics. But this moment of switching on, of illumination, of flooding the world with light, of dividing light and darkness, is a powerful image. It suggests a beginning, a difference between being and not being. The story inferred from cosmological evidence begins with a moment when time, space, matter and energy all happen at once.

But once you start asking a physicist, the story becomes more mysterious: there is no 'before' in this story, and there is no 'there' for this event to happen. Nor is there anything that humans could imagine as 'matter' because if the universe began, it must have begun from a virtual particle. In one of those say-it-quickly-and-it-seems-to-make-sense rubrics proposed by cosmic logicians, virtual particles exist in something called the quantum vacuum. We should pause here, and remind ourselves that we may be in not just unknown, but unknowable territory. No experiment can possibly replay the moment of creation. Nor can those of us who are not advanced mathematical physicists follow the reasoning of those who are. But advanced physics works from the assumption that what is not impossible, not forbidden by the laws of physics, must therefore be possible. That doesn't mean it must happen, just that it could happen. And one of those possibilities is the virtual particle. A virtual particle can pop into existence, and pop out again, in an interval called a Planck second – a subdivision of a second so fleeting it's almost impossible to imagine. No law of physics is violated: the energy implicit in this instant of existence is borrowed, and returned one Planck second later. The quantum vacuum – a phrase to describe a theoretical state made logical by quantum mechanical reasoning – is intricately involved in the moment of creation but you couldn't draw such a thing, you couldn't picture it.

Posters prepared by scientific organisations to illustrate for the ordinary reader the history of the universe from its first moment until now, almost always incorporate a brilliant explosion right at the beginning. But such an explosion – in the sense of a hot, a furiously hot, an unbelievably hot expansion of energy, a blast so astounding that, 13.7 billion years on, we can still take its temperature and record its radiance – would have been invisible: that is, it would not have involved any visible light. Astronomers and cosmologists have always assured me that for a sizeable chunk of the first million years no light would have been visible. The universe would have been too opaque, too 'optically thick', one said to me, for any telescope to detect. Light came later, and although the universe must have begun with a virtual particle, whatever that is, so did matter as we now know it: chunky, clunky stuff with mass we can measure and dimensions we can record.

Einstein's equation says that mass and energy are interchangeable, and long before the first tests of nuclear and thermonuclear weaponry, physicists knew that. A nuclear weapon turns small amounts of matter into pure and ever-so-destructive energy, but the same can happen in reverse. A particle accelerator can be thought of as a movie camera that records the play of matter and energy, and long before the construction of the Large Hadron Collider, researchers had observed the conversion of energy into matter. They recorded the sudden apparition, from seemingly out of the blue, of a positron and an electron, that is antimatter and matter, two discrete and observable lumps of stuff that – because they have opposite electromagnetic charges – describe a spiralling arc and then smash into each other, to become once again pure energy in the form of a gamma ray. So it is possible to conclude that all matter – not just electrons and protons but atoms and elements and compounds and polymers and flesh and blood and iron and soil – is simply condensed light, a blaze of brilliance, a burst of energy made solid and durable and static, at least for a time. It's a lovely idea, echoed or prefigured in various subtle or unsubtle ways in religion, philosophy, poetry and art.

There are unlovely problems with the same idea. One is that when radiation turns into matter it seems invariably to condense into equal quantities of matter and antimatter, two entities that are both perfectly real, but mutually destructive. In which case, why do we live in a universe seemingly composed almost entirely of matter, and in which we observe antimatter only in scientific experiments? And if matter and antimatter seek each other in acts of mutual destruction, why are we here? Where is all the antimatter we should be at risk from?

And if matter is just light brought to a halt and concentrated, if matter is just a distillation of light that drips from some kind of cosmological alembic, like brandy from a still, then why does matter have mass? Why isn't a bar of iron or a tub of lard as ethereal as a moonbeam? Both questions have putative answers: one possibility is that the matter–antimatter manufacture must have been slightly asymmetric, and our universe is the deposit, the residue left over after a whirlwind period of mutual annihilation that left only one form of matter standing. If so, continued observation of matter–antimatter creation should confirm that hypothesis. To test it, of course, you'd need to build a particle accelerator that could accelerate matter to almost the speed of light, that could deliver enormous levels of relativistic energy to each particle so that when it collided with another particle similarly accelerated, the impact would create a shatter of fragments, some of which would not be unlike those created in or just after the Big Bang that signalled the moment of creation. Some of these fragments would vanish into bursts of radiation and condense from radiation and destroy themselves again, and if you went on doing that experiment for years and years, you might be able to observe the proposed imbalance and finally say 'Aha! It is as we thought. In the great matter–antimatter conflict, our side wins by a whisker, matter overcomes antimatter: only just, but that's enough to explain the Milky Way, and Andromeda, and the Local Group, and everything we can so far detect.'

Of course, the same experiment has already tentatively answered the other question of why matter has mass; why something that began as light has condensed to a lump of something that can be weighed on a scale. There is a thing called the Higgs field, or a physical entity called the Higgs boson, that theoretically answers the question. Matter has mass because in what physicists portentously call the Very Early Universe, the Higgs field made it happen. It somehow slowed and detached and compressed packets of radiation, imposed a drag or anchor that turned a twinkle of light, a disembodied spirit, a free and lightning spark, into something more contained and earthy: a sprite into a sprat, a ray into a manta ray. What had been levity now answers to gravity. It is a very gratifying idea: we are all creatures of light.

The old idea of the soul – that disembodied, immaterial entity that informs identity, that delivers integrity, that registers responsibility, that responds to romance – is a notion not separable from light itself. The enduring idea of salvation and reunion with God in heaven is prefigured as a kind of merger of a quantum of light, or at least lightness, in the one luminous glory of the Eternity. Dante Alighieri puts it rather neatly. *The Divine Comedy* follows Dante's pilgrim through the circles of hell, and to the summit of the mountain of Purgatory, where beyond in heaven, his Beatrice is waiting to escort him to the celestial spheres, to confront the Lux Aeterna, the Divine Light itself. As a mere mortal, given a glimpse not just of Eternity, but of the Holy Trinity, the three-in-one light of Father, Son and Holy Ghost, Dante's pilgrim very reasonably complains that words cannot transmit the experience, but does his best:

> That light supreme, within its fathomless
> Clear substance, showed to me three spheres, which
> bare
> Three hues distinct, and occupied one space;
>
> The first mirrored the next, as though it were
> Rainbow from rainbow, and the third seemed flame
> Breathed equally from each of the first pair.

Dante strives to comprehend this mystery and, not surprisingly, when for a brief moment he does, he describes his moment of understanding as 'a flash'. In the last line of this great work, it all becomes clear: his will and his desire are turned by love: 'The love that moves the sun and the other stars.' God is light, good is light, ignorance is darkness, and to escape from the dark is to embrace knowledge – call it learning, science or scholarship – and beyond that, understanding. Which brings us back to Boethius, and his *Consolation of Philosophy*. My edition of Boethius dates from 1969, and I first read about him five years earlier, in a book by C.S. Lewis called *The Discarded Image*. In that year, NASA crash-landed a Ranger robot mission in the Sea of Tranquillity on the moon, but the television cameras on board failed to work: the mission ended in moondust and darkness. Months later that year, a second Ranger mission hit the moon's Mare Nubium at strolling speed – about 1.6 miles a second – and transmitted 4,300 images of the moon in its last seventeen minutes of descent. Soviet engineers despatched *Zond 1* to Venus, with a landing capsule aboard that should have recorded information about the nearest planet, its atmosphere and its surface. It arrived, but its communication system failed, and so that mission too ended in darkness and ignorance. If any of us were aware of these things at the time, we soon forgot them: 1964 was the year that four Liverpudlians delivered the global phenomenon of Beatlemania, the year Cassius Clay became world champion and changed his name to Muhammad Ali, it was the year that the Federation of Rhodesia and Nyasaland became independent Zambia, with Kenneth Kaunda as its first president, while the future of the region south of the Zambezi became ever more contentious. In apartheid South Africa, a court sentenced Nelson Mandela to imprisonment for terrorist offences. The actor Richard Burton married the actress Elizabeth Taylor for the first time. Britain elected a Labour government. The bloodshed in Vietnam continued, and worsened. I got married. My wife became pregnant. And a literary editor handed me a book to review. It was *The Discarded Image.*

I had already discovered C.S. Lewis: in the 1950s *The Screwtape Letters* had been hugely popular – the advice from a senior devil to an apprentice satanic emissary, even if readers didn't believe in Lucifer or Beelzebub. Many more, perhaps, had eagerly read his science-fiction trilogy, the one that began with *Out of the Silent Planet,* the story of a Cambridge academic called Ransom who is abducted, bundled into a spaceship and whisked off to Mars. It was fairly obvious, even to teenagers, that the trilogy was a parable and a commentary on sin and loss and salvation. We read it anyway.

The Discarded Image was Lewis the Critic, discussing the classical and medieval literary underpinnings for the Ptolemaic or pre-Copernican world view, the one in which Earth was the centre of creation, a nest of imperfection observed by the sublime citizens, elements or entities of the unchanging heavens beyond the moon. In it he airily took for granted that what he considered the three most important texts – the Bible, Ovid and Virgil – need not be discussed: 'most of my readers know them already'. He may not have been right even then, but he raced on to introduce Chalcidius, Macrobius, pseudo-Dionysius, Apuleius and the impact they had on Chaucer, Milton and Thomas Browne among others, and then devoted fifteen pages to Boethius, whose *Consolation of Philosophy* was for centuries 'one of the most influential books ever written in Latin'. It had been translated into English by Alfred the Great, by Geoffrey Chaucer and by Elizabeth I. 'Until about two hundred years ago it would, I think, have been hard to find an educated man in any European country who did not love it. To acquire a taste for it is almost to become naturalised in the Middle Ages,' Lewis said.

Boethius was a Christian Roman and minister to Theodoric the Visigoth, who at the time ruled Italy and, says Lewis, was in many ways a better ruler than many of the emperors had been. Theodoric, however, had his suspicions of Boethius, and had him arrested, detained and imprisoned. Boethius would later have ropes twisted 'round his head till his eyes dropped out' and then be finished off with a bludgeon in AD 524. In between his fall from power and his horrible end, he wrote *The*

Consolation of Philosophy, a work that has imagery, ideas and language that Lewis finds in Milton, Dante, Chaucer, Shakespeare and Fielding. Bertrand Russell, in his *History of Western Philosophy*, published in 1946, says that he cannot think of any other European man of learning so free from superstition and fanaticism during the two centuries before Boethius' death, nor the ten centuries after it. 'He would have been remarkable in any age; in the age in which he lived, he is utterly amazing.'

But Russell was not the only one to think so. Dante introduces Boethius, glancingly and not by name, in Canto X of *Paradiso*, as one of the twelve-strong circle of lights, all exponents of learning who have achieved sublime immortality. For more than a thousand years, 'many minds, not contemptible, found it nourishing', Lewis wrote of the *Consolation*. In the text, the visiting personification of Philosophy recites a set of verses to Boethius, and reminds him:

> This was the man who once was free
> To climb the sky with zeal devout
> To contemplate the crimson sun,
> The frozen fairness of the moon –
> Astronomer once used in joy
> To comprehend and to commune
> With planets on their wandering ways.

It cannot have been a very new idea, even in the world of Attila the Hun and Theodoric the Visigoth, the dismemberment of the Roman Empire, and the coming of the Dark Ages, to conflate knowledge and fulfilment with light, and ignorance with sorrow and darkness. It isn't hard to imagine that Boethius would have enjoyed some of the satisfactions of physics at least as much as his Neoplatonic contemplations. One of these is the way his intellectual successors have used light to extend humankind's knowledge and understanding of wider creation.

Boethius takes light for granted – we all do, those of us with eyes to see it – but there is a sense that the comprehending and communing astronomer within him really would have enjoyed the Hubble Space Telescope. The extraordinary and wonderful pictures taken by this orbiting instrument, this whirling eye-in-the-sky, have been a source of aesthetic and philosophical satisfaction and pleasure to tens of millions, and we all have our favourites. Mine is the Hubble Deep Field. It isn't the prettiest, the most dramatic or the even the clearest photograph, but more than any of them, it tells me something I had never truly understood, about light, and about darkness. This is the one that raises so many questions, about time, about matter, about the shape of the universe we live in, about the nature of seeing, and the nature of being. It also told me something about light, and about the way we register reality, both in a philosophical sense and in a practical way. But to make sense of the picture, you have to know a bit more about how it was taken, and a bit more about what light is.

It isn't enough to say that light is an electromagnetic wave, which radiates through a vacuum at an absolute speed. Or that optical light from the sun is white, or at least looks white, and that when fed through a prism it fractures into a suite of rainbow colours, which it could do only if it was a waveform, and not a particle, an atom or – the physicists' word – a photon, the smallest subdivision of light that is possible. It isn't enough to say that we call it optical light only because evolution has fitted our eyes with detector mechanisms that are best for certain wavelengths, represented by the run of colours from red to violet, if only because for all practical purposes, ordinary people in their ordinary business don't regard the wavelengths beyond the optical as useful for 'seeing' anything. But to a physicist, light is just light, a seamless stream of information that can be observed, interrogated, deployed and measured.

Light tells us about distance: if it is far away, it looks faint. If it is very far away, it may be barely visible. Even further, it won't be visible at all. As light radiates from a source, its brightness falls away as its energy diffuses with the square of the distance. The brightest stars are not necessarily brighter than the other stars in the immediate field of view. The logic of this argument is that stars that are even further away may be too faint, they may even be invisible. Of course, if we magnify the field of view with binoculars or a telescope, we can see more stars than we could with the naked eye, and the more light the telescope can collect, the better the view. So the number of stars in the night sky that we can see without a telescope or binoculars is numbered in thousands, but the galaxy of which our solar system is just a small detached property in an outer suburb contains maybe 100 billion stars, or 200 billion, or even 400 billion.

At least one of the stars we can see with a naked eye – until 1920 it was considered a smudge or cloud of star-making material – was called the Andromeda nebula, within the constellation Andromeda. After 1920, astronomers began to accept that it was in fact another galaxy altogether. It is now known to be a spiral galaxy of maybe a trillion stars, at least 2 million light years from us, but heading our way, on collision course with the Milky Way, some 4 million years from now. That's the unusual thing about Andromeda: it is coming towards us. Most galaxies are flying away from us, and we from them.

When the Hubble Space Telescope took the Deep Field image, the universe was known to contain maybe 100 billion galaxies, each with at least 100 billion stars. The universe should be flooded with light: there should be no darkness anywhere. But there is darkness, and to take the Deep Field image, the Hubble Space Telescope focused on one little speck of blackness in the night sky for ten days. Each of those terms needs its own gloss. First, the patch of sky was in the constellation Ursa Major. The little speck chosen was indeed little: 2.6 arc minutes. For those of us not used to thinking in arc minutes, imagine the vast entire night sky as composed of little grid squares each 2.6 arc minutes across. There would be 24 million of them. Hold a grain of sand at arm's length: it would obscure 2.6 arc minutes. This is a very small patch of sky. Now the ten days problem: the Hubble Space Telescope is in orbit. It moves. During ten days in December 1995, the telescope circled Earth 150 times, but kept its cameras pointing whenever possible towards this same spot. Photons of light from this small speck of blackness in the sky were collected by the telescope's enormous primary mirror, focused onto a smaller secondary mirror and passed to a set of scientific instruments that translated them into ones and zeros. This data was then beamed as radio signals to ground tracking stations and then to the Goddard Space Flight Center in Maryland and then to the Space Telescope Science Institute in Maryland where the intermittent, fleeting pinpoints of light were mapped onto a screen, a little at a time, until a picture began to develop. The result was a reminder that not only do we *not* know the extent of the universe, but perhaps we *cannot* know its extent; that beyond any worlds we can know, there may be more – many more than we could ever hope to see.

The Hubble Deep Field was a composite image, a montage of 342 exposures of photon capture by four broad-band filters each with a different wavelength, ranging from near ultraviolet to near infrared. In that prolonged-exposure plate of that dark spot in the sky, scientists counted between 1,500 and 2,000 galaxies, some of them 10 billion light years away, and therefore shining 10 billion years ago. If there were 1,500 to 2,000 galaxies in a speck of sky chosen at random, and there were 24 million such specks of sky to count, then that adds up to another 40 or 50 or 100 billion galaxies waiting to be discovered, simply by staring intently at one spot after another. In his book *Seeing and Believing* (2000), Richard Panek called that speck 'the hole that NASA drilled through the heavens'.

In 2003 and 2004, the Hubble astronomers drilled even deeper: they chose a spot in the constellation Fornax, and did the same again, only for longer, and called the outcome the Ultra Deep Field. Then they went further still and made Hubble stare (so to speak) at a patch of sky inside that ultra-deep field for twenty-three days, collecting all the photons, or particles of light as they arrived, and in 2012 published the Hubble Extreme Deep Field. This revealed

5,500 galaxies, some of them at distances of 13.2 billion light years. That is, the light collected by the telescope in the twenty-first century left those oldest stars 13.2 billion years ago, when the universe was young: perhaps 500 million years old.

Every picture tells a story, but the Hubble Deep Field and its successors tell so many stories one might need a whole book to do justice to them all. From just one picture taken through a pinhole in the sky, astronomers doubled their estimate of the number of galaxies in the universe. We already know why they are confident that they can do that: they can do it because the Copernican principle says what's true for this bit of the universe is also true for that bit of the universe. What you see will be roughly the same in any direction, and at any distance. The universe is full of galaxies, with huge spaces between the galaxies, and if you can get a really big picture of the whole bag of tricks, it will seem as though galaxies are somehow in groups, the way islands seem to fall into archipelagos and stars seem to fall into constellations. And just as between the galaxies there are vastnesses of emptiness, so between groups of galaxies there seem to be even vaster voids and then more groups of galaxies, seemingly without end.

The other important story is about time. We already know that the information delivered by the telescope is not about *now* but about *then*. When the batsman watches the ball fly from the bowler's fingers, he is already 'seeing' something that happened about one quarter of a second before he was able to see it, mostly because his optic nerve takes a moment to transmit signals to the visual cortex and then the brain itself requires time to process information that arrives at the speed of light, and turns it into a picture that delivers information about speed and trajectory, during which time the ball has already travelled a measurable distance. The batsman's information is always out of date. The astronomer's information is always old; what the astronomer's brain makes of it, too, is selective. Astronomers are humans: there is a danger that they will see what they want to see. The safeguard is that most of them understand that, and want their conclusions confirmed wherever possible by another observation, or an independent experiment, and if possible more than once. The batsman confirms his observations by hitting the ball all the way to the boundary: an even more convincing confirmation. Astronomers know that they are peering into the past: the wonder of physics is that they can tell, from the light they collect, how long that light has taken to get to them. When asked to explain how they can do this, they just say 'redshift'.

The lovely thing about light is that it is emitted by matter, and absorbed by matter. Matter isn't simply condensed light, it is forever involved in a gavotte, a square dance, a minuet with its partner, light. It is easy to assume that Albert Einstein must have received his 1921 Nobel Prize in Physics for his 1905 theory of special relativity, or perhaps his 1916 conclusions about general relativity, or perhaps the 1905 calculation of mass–energy equivalence (the famous $E = mc^2$), or all three. But although the Nobel citation mentioned his services to theoretical physics he was awarded it specifically for his study of the photo-electric effect. That is: shine a light at an object and some of that light will be absorbed by matter. Heat an object until it glows, and what you see will be light of very specific wavelengths being emitted by that matter. What those wavelengths, that frequency, that colour of light will be depends on the elements that have emitted that light: each element absorbs or emits quanta of light at very specific wavelengths, frequencies or energies. (Quantum mechanics starts from this phenomenon, but we'll save the fun of that for some other book, written by someone who knows more about it.)

Broadly, light carries the signature of the element that emits it. If light passes through a cloud of something, it picks up a signature of the elements it passes through, on its way to the telescope mirror, the camera or the eye. The eye can see this signature: one can tell that the tint of hydrogen light is different from that of sodium, or neon, but we see it imperfectly, as if from a broad brush rather than a fine pen. The spectroscope, however, can see everything perfectly: pass light through a spectroscope and the telltale absorption spectrum of the elements will be there, and astronomers know them as Fraunhofer lines. The lines that announce the presence of hydrogen, or lithium, or iron, are so precise and so characteristic that the first evidence of the existence of helium, and even its name, came from an analysis of sunlight: *helios* is the Greek word for the sun.

Another key factor about light is that it travels in waves. What is true of sound waves is also true of light waves. Light, like sound, becomes more energetic as its wavelengths become shorter and its frequency higher. Just as a high-pitched noise becomes painful, and then at even higher frequencies, inaudible, so high-frequency light is more energetic: violet light is just fine, ultraviolet light is invisible, and potentially damaging. And we know that the Doppler effect predicts that sound waves emitted by a source travelling towards the listener will not seem the same as the waves emitted as the source is alongside the listener and they will be different again as the source speeds away from the listener. So light from a star heading towards Earth would show a squashing of the wavelengths: they would seem to have more energy or a higher frequency. There would be a shift towards the blue. Light from a star rushing away from Earth would arrive with its wavelengths elongated, its energy reduced, its frequency lower: a redshift. The speed of light is absolute: it leaves a star at 300,000 kilometres per second, and arrives at Earth at the same speed, whether the star is rushing away from us or rushing towards us. The only change is in the frequency, or to use another test, the colour. Since stars can have different colours to begin with, this blueness or redness would of itself tell us nothing. But the Fraunhofer lines – the little sequence of lines that serve as a signature, that pronounce, with firmness and emphasis, the presence of hydrogen, or lithium, or helium – would appear in unexpected places in the spectrum: because the relationship between these telltale shadows, the Fraunhofer lines, remains the same, astronomers can be sure that they are looking at a redshift. And the further the shift in the signature lines, the greater the redshift will have been. The greater the redshift, the faster the source that emitted the light is receding. That is, essentially, how astronomers know that galaxies are (on the whole) receding from each other, as if the universe was expanding. That's the simple version.

Redshift isn't actually that straightforward – a gravitational field can affect light too, so cosmologists must distinguish gravitational redshift from the cosmological kind – but we'll stay with the simple version, because it tells us something: if the laws of physics are the same everywhere in the observable universe then you can make some reasonable assumptions about anything you can see. And one of the things you can say is that the galaxies pictured in Hubble's famous Deep Field are not just very faint and very far away, but they are also receding from us very quickly. The ones receding at the fastest rate are the ones that are farthest away. Some redshift speeds observed by earthbound astronomers seemingly approach the speed of light. That means that to a hypothetical observer on one of these galaxies, this one, the Milky Way, would also be receding at almost the speed of light. As far as we are concerned, we are right here and going nowhere, so that's what it must seem like on those galaxies 10 billion light years away, and 10 billion years ago. This is only possible because space itself is expanding. Space is on the move, and taking matter with it. If the light that left galaxies 10 billion years ago shows that they were already parting from us at speeds approaching the speed of light, then it follows that in the intervening 10 billion years, they have travelled even further, at an ever faster rate, so that at some point, the light-emitters that we think we have seen could be already over the horizon. The light they emitted last year, or a million years ago, or a billion years ago, may never reach us, because as far as we are concerned, that distant, seemingly motionless bunch of galaxies would have separated from us at velocities exceeding the speed of light in a vacuum. And we can tell ourselves self-evident truths about long ago and faraway – preposterously long ago and unimaginably far away – thanks to an intellectual revolution that began four centuries ago. We can be sure, thanks to the principles of the telescope as devised by Newton and Herschel and others; and thanks to some basic mathematics and reasoning by James Clerk Maxwell and Einstein and his contemporaries, and thanks to the mechanics of rocketry advanced sufficiently to put a telescope beyond the atmosphere that puts the twinkle into the starlight, that removes a certain amount of important detail in the process.

As Douglas Adams so cheerfully put it in his 1978 BBC radio series *The Hitchhiker's Guide to the Galaxy*: 'Space is big. You just won't believe how vastly, hugely, mind-bogglingly big it is.' And of course, Adams was concerned only with a single galaxy and corresponding galactic distances. There is a certain comedy in the information delivered by Hubble Deep Field, which is that, after a certain level, size becomes meaningless. What does it mean to decide that the observable universe contains 40 billion galaxies, or 100 billion galaxies, and then derive new information that forces a rethink: no, make that 200 billion, or 400 billion? It certainly doesn't make us any more insignificant than we already are.

If our planet was an insignificant object within the solar

system – which it is – and our parent star the sun is only a midlife main-sequence star in a galaxy that is home to 100 billion stars, and those stars add up to only a fraction of the real mass in the galaxy, because most of this mass is dark matter, something so far undetectable, then our significance is already clear: as lumps of matter, we don't count. If we weren't here, the rest of the galaxy would never know. The disappearance of planet Earth and its creatures would make no difference to the solar system. Beyond the solar system, on the stars nearest to us – those a mere four light years away, or forty light years, or even 400 light years – no other hypothetical observers using technologies that match ours could ever know that we were once here. The galaxy is 100,000 light years across.

To early twentieth-century observers our status within the universe was without significance of any kind. They thought that the galaxy they observed was in fact the whole universe. Now, the size of the universe is anybody's guess. It had a beginning. That seems to suggest the universe could be finite. The logic of observations says that if the universe began 13.8 billion years ago, then the observable universe would be limited to a distance 13.8 billion years in any direction. That gives us a spherical universe notionally perhaps 28 billion light years in diameter. But things we could see that emitted their light 13.8 billion years ago have gone on receding from us in those 13.8 billion years, so that gives us a universe at least 46 billion light years in diameter. And none of that takes into account a strange episode many physicists believe happened within the first millionth of a second of creation. They call this event cosmic inflation. It is not the same as the expansion of space that astronomers can measure, and calculate. This event would have bestowed on the universe the properties it has, the properties that mean that what is true for us must also be true everywhere, and the sameness we observe from our limb of the galaxy must be true of all galaxies, including the ones we could never see. In this event, somewhere in the first millisecond, the universe-to-be suddenly inflated from the dimensions of a subatomic speck to some enormous scale, and then this period of inflation stopped, leaving a universe – one that may have been the size of a beachball, or may have been light years across – to go on expanding at the rate it now expands, for a subsequent 13.8 billion years, and although this puffing up, this inflation, really doesn't make sense, in another way it does. It explains everything. Or would, if we could understand it.

V

The Distance of Darkness

What are we to make of these discoveries? What would a listener at one of Plato's symposiums, or a senior civil servant observing the ruin of the Roman Empire, or an Elizabethan adventurer or a Florentine poet make of all this? My guess is that, after an initial gulp or two and a bit of reflection, the reaction would be the same as ours: a mix of acceptance, bewilderment and disbelief. The acceptance is possible because, *prima facie* and without the ability to test the evidence, the story told by cosmological science is no more difficult or easy to accept than the one that provided the uncertain core of Christian belief for 2,000 years: that God made the world in six days, and on the seventh he rested, and man and woman were among the initial acts of Creation.

I call it the 'uncertain' core of Christian belief because it is quite clear that St Augustine of Hippo had his doubts about the literal nature of such a story: he points out, in his book *The City of God*, written sometime before AD 450, that the Bible never lies, and acknowledges the days of Creation, but then adds 'What kind of days these are is difficult or even impossible for us to imagine, to say nothing of describing them.' Augustine points out, very reasonably, that we count days as periods marked by morning and evening, by sunrise and sunset, 'whereas those first three days passed without sun, which was made, we are told, on the fourth day'. When the light was made, he listed the nature of that light as one of a number of questions, along with the nature of morning and evening, as 'beyond the scope of our sensible experience'. He also addresses questions not mentioned in the Genesis story, one of which was whether time had always existed, along with space, when God made the world. He concluded that if you did place God in space that had always existed, why did he make the world here, rather than over there? And that if time was infinite it was fair to ask the question: why create the world now, and not then, or at some other time? 'We are not to think about infinite time before the world, any more than infinite space outside it,' he says. 'As there was no time before it, so there is no space outside it.'

The Enlightenment happened, and we who were born afterwards sometimes think of ourselves as Enlightened, but it is pretty hard to read translations of Augustine, or Seneca, or Pliny the Younger, or Suetonius or Lucretius, and still think that they inhabited some kind of philosophical darkness, that they were incurious, or ignorant, or bigoted, or afraid to think for themselves, or that they would not be interested in the same questions that exercise us. The language of the canonical Scriptures has always been a problem, especially for thinkers who want to square the Bible with the non-biblical world.

One of my favourite touchstone people for such matters is the Reverend Samuel Kinns, fellow of the Royal Astronomical Society, who in 1882 wrote a rueful, anti-Darwinian tract called *Moses and Geology, or The Harmony of the Bible with Science*. Kinns had no problems with the idea of 'day' as an indeterminate period that could encompass aeons, nor did he have any problems squaring the cosmogony of Moses – the supposed hand that inscribed the first five books of the Old Testament – and the events of the first six days of Creation with the story as told by nineteenth-century astronomy, archaeology, palaeontology and geology. He would choose a geological period, or a palaeontological event, and match it with an account in Genesis. He reasoned that the story told by science and the story told by Moses matched, in the order of time, in both the geological record and the biblical text. The chances of his fifteen selected instances being transcribed in their proper order in the Bible without divine inspiration, he reasons, would be a billion to one. Here is a sample of his reasoning:

> VI. – Science: Next succeeded the lowest class of Phaenogams, or flowering plants called Gymnosperms,

from having naked seeds, such as the Conifers. Dana mentions coniferous wood found in the lower Devonian.

Moses: *'The herb yielding seed.'*

VII. – Science: These were followed by a higher class of Phaenogams, or flowering plants, bearing a low order of fruit, found in the Middle Devonian and Carboniferous strata.

Moses: *'And the fruit tree yielding fruit.'*

The Higher order of fruit trees appeared when 'God planted a garden' later on.

VIII. – Science: The Earth until after the Carboniferous period was evidently surrounded with much vapour, and an equable climate prevailed all over its surface; afterwards these mists subsided, and then the direct rays of the sun caused the Seasons.

Moses: *'And God said, let there be lights in the firmament of heaven, and let them be for signs and for seasons.'*

There is a profound logical problem with any attempt to parallel scientific discovery with Holy Scripture. The biggest of these is that science sometimes gets things wrong, but the Bible cannot be wrong. So if the two square, all well and good. But science considers all observations and their interpretation as provisional. A working theory is only as good as the data that supports it. Science operates on the working principle that researchers can get things wrong, and sometimes do: at some point error is corrected, and the parallel you thought might be there becomes less apparent, and the exercise becomes pointless. One feels that Boethius would have seen that logical weakness. In his time Kinns was considered to be intelligent and thoughtful. Yet on the evidence of some of his reasoning it is hard to see today why anyone should have thought so. By comparison, Boethius, St Augustine and many of the Roman historians and chroniclers sound no more wrong or wrongly informed than Thomas Henry Huxley or Alfred Russel Wallace or Charles Darwin would have been in their time.

We look back at the past as if from the summit of a cairn: a pile of stones of understanding carefully placed there by the people who went before, and replaced by generations that observed and corrected any perceived errors. So we see more, because we have the advantage of height conferred by the understanding delivered over 500 or 1,500 years of previous inquiry. We are no cleverer; but we have more and, we hope, better information. In one sense any physics A-level pupil knows more science than Isaac

Newton did. But we know that Newton was cleverer than almost anybody who went before, or after.

It isn't hard to see that Augustine of Hippo, who did not have the benefit of geological strata, fossils, high-resolution telescopes or any clear idea of the nature of light, was so much wiser and shrewder and perhaps even in some ways better informed than Dr Kinns. And, one suspects, Boethius too would be a match for any of us now. So they would respond as we do, with delight, to the idea of an expanding creation, of a universe crammed with matter – black holes, neutron stars, main-sequence stars, red giants, white dwarfs, brown dwarfs, planets, asteroids, meteors, comets, comet dusts, gas clouds, the debris of supernovae, clouds of chemicals, vast mists and nebulas of hydrogen – and informed by light, but still somehow almost entirely empty, and getting seemingly emptier, as the galaxies rush away from each other at speeds that increase with distance. They would ask detailed questions about time, and space, and matter, and light. It doesn't mean that they would reject the Scriptures. It is perfectly possible to believe in God the Creator, and regard the Scriptures as the best available anecdotal evidence, and still wonder about the fine detail and the temporal sequence of creation itself: how the cosmos unfolded from a moment before which there was no time, and in a space before which there was no space, to be populated by creatures that, ultimately, had no begetters.

One of those questions asked would be: what is space? If it wasn't there before, it must be something rather than nothing. If it can expand, is it a fabric, and how far, ultimately, can it stretch? And what does it mean to say, with Einstein, that matter distorts space, and things fall to Earth because space dictates how matter moves? Is that ultimately any easier or harder to understand than the Newtonian idea of mass having attraction, of exerting a pull, of imposing action at a distance? Or Aristotle's proposition, in his work *Physics*, that things fall because they have an innate property called gravity, and soar because they have levity? Aristotle also argued that things fall at a velocity proportional to their weight or mass: heavy things fall faster than lighter things. He qualified it by saying that the velocity is also inversely proportional to the density of the medium they fall through. Even though Galileo demonstrated that this was not so, and even though two *Apollo 15* astronauts on the surface of the moon filmed a hammer and a feather hitting the surface at the same time, there is a sense in which, when we think about such abstractions at all, we still think of heavy things falling faster than light things. They certainly fall harder. Once again, we think of ourselves as better informed, but we cling to old notions.

We know Earth is a sphere, but we also speak of the four corners of the earth, or the ends of the earth, as if it was flat. We think of medieval ideas as dated and wrong, but we preserve them in our language, our metaphors and imagery. And medieval philosophers were as just as clever and as thoughtful as we think ourselves to be, but within a different intellectual framework. Ours is defined by Einstein and Newton, theirs by Aristotle. Indeed, we know a little about what Boethius may have thought of Aristotle's ideas of gravity, but only by inference: he set himself the never-completed task of translating Aristotle's works into Latin. But that doesn't mean that medieval thinkers got things wrong. Sometimes they got things right without necessarily formulating or applying what would now be called a scientific principle. Some years ago, an Italian physicist wrote to the journal *Nature* to point out that Dante Alighieri had anticipated by 300 years Galileo's principle of invariance when, in *Inferno*, he described making, in utter darkness on the back of the monster Geryon, a further descent into hell, a slow descent he would not be aware of, but for the 'sight of the beast appalling':

> And on he goes, swimming and swimming slow,
> Round and down, though I only know it by feeling
> The wind come up and beat on my face from below.

In effect, but for the wind on his face, he might be at rest. In 1632, says Leonardo Ricci of the physics department of Trento University, Galileo Galilei described his experiences on a tall ship and used them to explore the principle of invariance. This is one of the pillars of modern science. It says that the laws of motion are the same in all inertial frames. This means that if you were below deck on a ship sailing steadily at a constant speed you wouldn't be aware that you were moving. Less obviously, someone sitting in an armchair with this book is moving through the void at thirty kilometres a second, racing along aboard spaceship Earth on its year-long journey round the sun. Dante intuitively grasped a fundamental principle of science, despite his lack of data and technology. 'Still, it seems he was well ahead of his time with regard to views about the laws of nature held in the Middle Ages,' says Dr Ricci.

But this is not the only example of prescience from Dante. He seems to have understood another scientific principle. *The Divine Comedy* is of course a religious allegory, a story of the soul's glimpse of the afterlife, with its promise of damnation in hell, suffering in purgatory and salvation in paradise, but it is more than that. Dante has also created a world, too: one with topography, direction, orientation, co-ordinates, space-time, the lot. It is a spherical world, a planet, with places and their antipodes and with day in one location and night in the other. It is also a journey to the centre of the earth. In Canto XXIV of *Inferno* he is just about at the bottom of the Pit, and there his guide Virgil leads him to where the traitors Judas and Brutus and

Cassius are forever being devoured by Satan, who is himself stuck fast in the ice. Dante is a bit frightened by what happens next – they must climb down the body of Satan himself, as if it was a ladder to the next stage – so he does what a child might do, he clings on to Virgil and shuts his eyes. He senses Virgil feeling his way down Satan's shaggy flanks, hanging on to matted hair and crusts of frozen rime:

> And when we had come to where the huge thigh bone
> rides in its socket at the haunch's swell
> my guide, with labour and great exertion
>
> turned to where his feet had been, and fell
> to hoisting himself up upon the hair
> So that I thought us mounting back to Hell.

At one point Virgil has to cling to the satanic cliff-face like a crab, and reverse direction. Dante thinks they are going back the way they came, but after a while they pass through the rocky vent, or chimney, at the base of hell and take a breather, and Dante opens his eyes,

> thinking to see the top
> of Lucifer, as I had left him last
> but only saw his great legs sticking up.

And I stood dumbfounded and aghast
let those thick-witted gentry judge and say
who do not see what point it was I'd passed.

This was 350 years before Sir Isaac Newton outlined the theory of gravity, and made the world understand that you had to think of gravitational force as coming from a point source, at the centre of the mass, from which every direction will be upwards. Dante didn't invent a theory of gravity in his poem: he was just solving a problem of topography. But it's pretty obvious that, were Dante, or Boethius or Augustine, in some absurdist time-travel fantasia, to find themselves in a twenty-first-century university lecture hall, they'd soon be delighted by some of the things they heard, especially cosmic inflation, and the debate about time: St Augustine, in particular, would enjoy the logical conclusion that time and space are not separable, and should be thought of as space-time. Augustine proposed that the distinction between eternity and time was that, without motion and change, there is no time, while in eternity there is no change. He would have loved the Planck second.

A Planck second is the shortest period of time physicists can logically work with. If you wrote it out in the kind of notation that scientists use for quick, order of magnitude calculations, it would be 10^{-43} of a second. Another way of putting it would be to say that measurement would involve a decimal point and forty-three zeros before you get to the digit. Or you could call it 10 million trillion trillion trillionths of a second. And that is only a handy approximation: the true figure is more precise than that, and actually a bit less than 10^{-43} but even then, I am assured, still an approximation. It is arrived at by making an equation of three constants: the speed of light in a vacuum, the gravitational constant, and an entity called the Planck constant, and it takes its name from the reasoning in 1899 of one of the giants of quantum physics, Max Planck.

What matters about Planck time is that there is no way to work out what happened before the first tick of this tiny interval. When the space-time begins, the clock has already registered this infinitesimal tick. There was no before: before we can possibly have known that it was about to happen, the universe had arrived, and with it, the past. This Planck second measures the rate at which things happen in the quantum world, and there are a lot of them. It is a fatuity to say that there are more of these units in a single second than there are seconds in the entire 13.8-billion-year history of the universe: if the universe was a trillion years old there would still be vastly more. From our viewpoint – we have a stimulus response time of about a quarter to a fifth of a second, so the speed of thought isn't actually all that fast – such subdivisions of the second are meaningless. But they are not meaningless to the subatomic particles of which the atoms from which everything we can see and touch are ultimately made. The things that happen at this seemingly ultimate subdivision of creation happen on scales so fleeting we can barely imagine them, even if we can write them down as units, and even measure some of their coarser multiples. And it is in these tiny units of time that particle physicists at places like CERN in Geneva try to calculate. And one of the events that could only be measured in these units is cosmic inflation. This is an idea that – I was assured when I first heard of it – might seem mad, but not to particle physicists.

Some phrases favoured by physicists are shorthand for our inexplicable beginning. One favoured but bewildering phrase is 'the initial singularity' to describe a one-dimensional point that must have contained time, space, matter and all potential stars and planets, to burst forth in what has become another favoured phrase, 'the Big Bang'. The imagery sums up in two words, but can hardly explain, the unfolding of time, space and substance in one dense, hot blast deep inside the first second of creation. Cosmic inflation is a proposed interval within that short space of time. It has been discussed so often in journals, books and television programmes that it is easy to assume that it definitely happened. It remains a conjecture, yet a conjecture that so far has proved consistent not just with observation and evidence, but also with predictions of evidence later to emerge. It explains several otherwise puzzling properties of the universe, so cosmologists have examined the idea and reformulated it in several versions. But the big central idea goes like this: in or around the first Planck seconds, the universe existed only as a vacuum. That doesn't mean it wasn't something: to a mathematical physicist, a vacuum is not nothing. It is already space, something with energy, density and pressure. Theorists play with ideas like this, and one of the ideas is a theory called a false vacuum. That is just a label for a special state in which a vacuum could exist. It may be 'just an idea' but there is no proviso, no veto, no caveat that says it is impossible. And if something can exist in theory, physicists have learned to accept that maybe it exists in fact.

Antimatter emerged as a theoretical possibility, until it was discovered. Time travel remains (to physicists) an unresolved puzzle: time is an aspect of space-time. That means it is a dimension, so could you travel within it? If not, can you establish logically why not? What mathematical reasoning says you cannot? Physicists, in particular, have learned to trust their mathematics, which is why they like the idea of a false vacuum. And they especially like one property of this false vacuum: in it, the force we think of as ordinarily an attractive one, gravity, becomes utterly repulsive. That is, it repels, it drives away. Things don't implode, they explode. We live in a low-energy vacuum right now, and have done since sometime after the first ticking second of time, but at some point in that first infinitesimal fraction of a second, the universe was a high-energy false vacuum. And it launched a superfast, exponential expansion. We are talking here about some tiny speck of notional space, but how could you call it space when it is measured in units of length so small we have no words for them?

A hydrogen atom is the smallest thing we experience. A hydrogen atom is almost entirely empty space. The hard core of a hydrogen atom is a proton. The dimension of this ultimately small length is about 100 million trillionth of the diameter of a single proton. Theorists call it quantum foam. In inflation theory, this tiny accident, this false vacuum crammed with antigravity, went into overdrive, and started to double, and continued to double exponentially. This doubling time is very short: a million Planck seconds or so, but that still adds up only to 10^{-37} of a second. And it went on doubling in size. In a few hundred of those units of doubling time, this speck of quantum foam, this notional entity crammed with antigravity, this speculative contradiction of reality, had become enormous: it had grown by the power of 10^{80}, or even more. Nobody knows exactly how much it grew. What theorists do know is that the process stopped. The process stopped because, theorists say, a false vacuum is an unstable entity. When it stopped is not known, but the reasoning of cosmic inflation theory is that it was over within the kind of timescale in which Planck seconds make sense. To a cosmic physicist, the time at which the false vacuum became something that could be considered as space-time and energy as we understand it would be measured at 10^{-31} of a second. Beside this interval, a twinkling of an eye would seem like an eternity. We don't have words for such brevities.

At the end of this notional time period, the universe had arrived: repulsive gravity had become energy, a superhot, superdense fireball, a concentration of unbelievable energy from which everything we on Earth now see and everything we cannot see and may never see will gradually condense: galaxies, and galaxies beyond those galaxies, and beyond them still more galaxies. Each of these galaxies will be visible as light but will also be embedded in some other matter with mass that can be calculated, matter that does not radiate, or interact. Theorists call it cold dark matter. It is there, but so far nobody knows what it is. And the same episode has left another legacy: empty space that expands at a rate that astronomers can calculate and argue about, and time that can be systematised in seconds, seasons, lunar months, solar or sidereal years, and then aeons and epochs and eras.

I was lucky enough, one night in Boston, Massachusetts, decades ago, to sit next to the man whose name is most often linked with the theory of cosmic inflation. I already knew that it existed as a theory, and that people talked about it, and I had already heard his summary of the idea: that the universe was the ultimate free lunch, because the entire panoply of everything we see and everything we don't was conjured up from nothing, or possibly a mass of something ever so small. By this reasoning, the universe is a borrowed object, and we are all living on borrowed time. We don't know who did the lending, or what bank the energy that drives the universe was borrowed from, but we do know that the sum of the universal assets and debts of the existing universe is round about zero. And there I was, sitting next to the man who could explain it all, always supposing I could understand what he had to tell me.

At the time, cosmic inflation sounded like just another conjecture about the prehistory of the universe. The cosmic bestiary at the time was full of things that theorists proposed must have existed in the Very Early Universe, or might still exist somewhere as survivors of this strange era, but had not yet been found. In the 1970s, we started hearing a lot about black holes which might or might not really exist, but which logic said might exist. In the 1980s, I had already been mesmerised by an account of a phenomenon called cosmic string. This would be a one-dimensional length of something almost infinitely thin that either stretches to infinity or forms loops a million light years long, and moves at almost the speed of light. And this string had another weird property: it was massive. One inch of the stuff would weigh about 10 million billion tons, and, said its intellectual begetter, were it to pass through someone or something, that person or thing would implode at 10,000 miles an hour to a similarly dense, massive point.

Cosmic string didn't necessarily exist, but the logic of a certain chain of speculation said that it might. Think of a flaw in a crystal: what is it but a fine, elongated empty space in something very hard and durable? So what would a fissure in the vacuum look like? It would look perhaps like cosmic string: think of it as a random flaw in creation, a little potential system failure waiting to announce itself while the universe is still running itself in, like a defect in a new car. Something with such a mass might be one of the theoretical candidates for the dark matter (there were many possible candidates, some with descriptive catch-all names such as wimps – weakly interacting massive particles – or machos – massive compact halo objects) that had to make up the missing mass that makes up most of the universe.

Cosmic string had some satisfying properties – it could, for instance, help explain why the galaxies formed in the way that they did, because another puzzle to be addressed was: why is the universe like it is at all, in big lumps with big spaces in between? Why isn't all the matter smeared all over the place? It would be possible to employ cosmic string as the invisible, gathering spirit that triggered the drift of objects closer to each other before dancing nimbly on. Cosmic string had another useful property: if there was any in the vicinity, you'd certainly know about it – stars and planets would start to disappear.

I quote cosmic string as an instance of such entities permitted by mathematical physics, but so far not observed. The begetter of cosmic inflation, the American physicist Alan Guth whom I had sat next to in Boston, had been puzzling in the early 1980s about another little riddle of the mysterious Very Early Universe. Whatever happened in the Very Early Universe, according to the reasoning of the time, should have delivered some very unusual particles, and along with cosmic string, there was the magnetic monopole. This is something outside human experience, and it defies the ordinary imagination. You have to imagine a lodestone that points to the North Pole but not to the South Pole at the same time. In the theoretical logic that played with the sequences of action that went from the first Planck second to the universe as it is now, things happen, and the theories said that one of those things should be the creation of magnetic monopoles, and lots of them too. Magnetic monopoles would be very hard to see: in dimension, they would be about 100,000 billionth the diameter of the nucleus of a proton. You'd know they were there, however, because of their mass, which would be 10 billion billion times the mass of a proton. The universe should, in theory, be littered with these enigmatic leftovers from creation, even after all the billions of years of the steady expansion of space. But none had ever been spotted. Guth and a colleague picked up the puzzle to see if the apparent invisibility of an entirely theoretical entity could be answered by an entirely hypothetical event. And the event they chose was supercooling.

Supercooling changes things from one state to another. A volcano can pump superheated molten rock into the air and in the ordinary way it will cool to rock again, maybe with vacuoles left by bubbles of gas, but it will still be crystalline rock. Blast such a piece of hot molten rock into cold water and it supercools to become a glass, such as obsidian. That seemed like a possible explanation, and then they thought of another question: what would supercooling do to the rate of expansion of the tiny, nascent universe? It was Guth who did the sums: he told me he went home and did the mathematics late one night and realised that the impact would be radical: instead of expanding at the rate we know now (which is pretty fast and gets steadily faster the further away the receding object is) the baby universe went like a bomb: in an impossibly small fragment of time it inflated 10^{90} times its own volume and became a fireball.

The piquant aspect of this discovery was that, as Guth told me at the time, it didn't seem necessarily an important discovery. It was just a hypothesis to explain the non-evidence of another hypothesis. In this hypothetically inflating universe, however many magnetic monopoles there were, they would separate from each other at such speed and to such vast distances that they would seem to have been diluted out of existence. The chances of seeing one would then become vanishingly small. So that was a possible solution. It took him a while – and conversations with cosmologists in other institutions – to realise that cosmic inflation theory did have other qualities that seriously recommended itself to the wider community. It looked like a plausible answer to huge and intractable problems. One of these is: why is the universe seemingly the same in all directions? This is called the horizon problem.

Room temperature is an idea you can live with, but room temperature is achieved over time: cool air that comes in from the open window and warm air that rises from a radiator need time to achieve an equilibrium. If the room is very big this might take a very long time. But a universe that began with a hot big bang and expanded in all directions wouldn't be like that at all because the components wouldn't have had time to mix and reach equivalent temperatures. Cosmic inflation solved that problem in a puff: whatever the state of the universe in its tiny initial state would have been replicated in this astonishing moment of inflation: everywhere in the universe would be the same, including the tiny inhomogeneities that permitted the formation of galaxies according to the same pattern everywhere.

The other problem was the flatness of the universe. Why did it look as if the universe was so finely balanced between universal expansion and a potential collapse? When people talk about the 'curvature' of the universe, they propose two possibilities: one invites you to imagine that the universe is curved like a sphere, and the other that it might be curved like the bottom of a bowl. If the former, perhaps gravitational forces would one day slow the rate of cosmic expansion, bring it to a halt and then a cosmic collapse, in which space, time and matter all ended in a Big Crunch? If the latter, would the universe expand forever, leaving every galaxy alone, growing ever colder in eternal darkness? It seemed, especially then, in the 1980s, as though the gravitational force that imposes a drag on all matter, and the pressure to expand and go on expanding, were balanced with such exquisite precision that it left the cosmologists unable to tell: either outcome might be possible. But why would these two competing forces – the pressure to expand and the gravitational tug to contract – seem to be almost precisely balanced? People called this, for shorthand, the flatness problem. And the theory of cosmic inflation disposes of this puzzle too. After a period of inflation on the scale proposed by Guth and others who took up the idea, the question would start to lose its meaning: flat or not, our bit of the universe would certainly look flat, just as planet Earth looks flat to those of us walking around on it. So inflation theory joined the catalogue of potential answers to the great question: why is everything like it is? Why is there something rather than nothing, and why does this something have this property and not that one?

The delight many of us take in these questions is matched by the delight we can derive from the potentially plausible answers. It doesn't mean that these answers must be right. Even if they turned out to be wrong, they'd be fascinating. The solutions to these puzzles don't deliver any obvious, immediate or apparent reward, but that is not why we are fascinated. Like Boethius, we can lose ourselves in them. They deliver satisfaction. We could dismiss this as a distraction from the real world, but that would be illogical and perverse. The whole point of such research is to understand the real world. As the personification of Philosophy explains, somewhat crisply, to Boethius, he has no business feeling sorry for himself because Fortune has failed him. That's Fortune's job, to go around randomly favouring people, and then deserting them. What really matters, she tells him, is what is inside his head. His destination is true happiness. 'Your mind dreams of it,' she says, 'but your sight is clouded by shadows of happiness and cannot see reality.' Earthly preoccupations, such as wealth, power, celebrity, good looks and so on cannot guarantee happiness. And why be the slave of that worthless and brittle master, the human body? 'Look up at the vault of heaven: see the strength of its foundation and the speed of its movement, and stop admiring the things that are worthless. Yet the heavens are less wonderful for their foundation and speed than for the order that rules them.'

And that indeed may be what lies at the heart of this fascination with physics, with the circumstances of creation and its process, and the strange revelations delivered by what it is tempting to call the high priesthood of physics: the sense that there is order and reason behind even the maddest things. Space, in the sense of that colossal emptiness in which matter falls, the space that in our immediate solar system is lightly clouded with dust, gas, pebbles, micrometeorites, asteroids, comets, minor planets, innumerable companion moons and big predictable partners that orbit the sun, that is flooded with sunlight and blasted by fiery particles accelerated from the surface of the sun, is not emptiness. It is a something. It has composition. This composition looks as if it were nothing, but, if it truly were nothing, then it is not possible to imagine how it can be distorted by a massive planet. Nor is it possible to imagine it being temporarily disturbed by strange events in the faraway cosmos.

In 1916 Einstein unfolded a new vision of space-time, matter and gravitational forces that has been repeatedly confirmed by experiment and observation, and in 2016, confirmed by an observation some of us had not expected ever to see. The evidence of gravitational waves rippling through space, evidence of some unimaginable collision between two black holes – invisible repositories that entomb the fragments of what were once twenty or thirty solar masses – that began to sashay around each other in ever decreasing circles and at speeds approaching the speed of light until they collided, and became one. And the sheer violence of this consummation, this torque and twist and wrenching of the local space around them, sent ripples out from the scene, to travel in all directions, getting weaker and weaker as they diffused with time, to travel for at least a billion years before they hit two detectors on planet Earth, to ring a bell, to leave a little record of their passing, a little chirp, a pattern of spatial vibrations from which forensic physicists could reconstruct the death throes far away and long ago. It is as if a sheet of space had been shaken like a blanket, and this shaking had continued all the way along the fabric, getting smaller and smaller, but still carrying that memory of a long-ago, faraway, epic convulsion. Perhaps that is the wrong metaphor: it suggests space is two-dimensional. Perhaps the metaphor should be that of an earthquake, a spacequake, a tremor or violence or fault movement in space-time. Earthquake waves travel at predictable speeds – predictable depending on the nature and density of the rock through which they are travelling but tend to hit seismic detectors at two or more kilometres a second, and hit them everywhere, at different times, depending on the distance and the density of the medium. An earthquake off Japan will hit the seismometers in Tokyo first, but detectors in London, or Istanbul, or Mexico City will all, before too long, and at a more or less predictable set of times, receive the same signal, and be aware of the cataclysm, of the deep and potentially destructive disturbance in somebody's vicinity.

What matters here is not so much the spacequake, but the fact that somebody predicted that such a thing could happen, and somebody else started trying to find ways to detect it when it did happen, and that sooner or later physicists began to devote ever greater sums of money to devising ever more sensitive instruments that might record its happening. When Dame Philosophy lectures Boethius to consider the order that rules the heavens, she might have had gravitational waves in mind.

I first heard of gravitational waves more than forty years ago, during my very first, diffident ventures into science reporting, and so long ago that I can no longer pinpoint the moment. It may have been from a more experienced colleague explaining that in addition to optical and radio astronomy, there were also infrared and ultraviolet telescopes, in those days mounted on rockets for short observations above the atmosphere, and – he said – gravity-wave astronomy. At which point he told me about Joseph Weber, an American physicist who since 1958 had been trying to detect gravity waves using massive aluminium cylinders isolated from all other disturbances, in his laboratory.

I do not claim to have given the subject much thought at the time. The notion that space could be distorted by violent events in the faraway galaxies had no great purchase during the Cold War: we tended to think more about the thermonuclear demolition of global civilisation triggered by some lunatic posturing in the world of geopolitics. But on 12 February 2016, on one of my last assignments as a science reporter for the *Guardian*, I had to retrieve the memory: a team of physicists had announced the detection of gravitational waves. They had confirmed the detection using not one but two instruments of exquisite sensitivity, with which they had recorded the collision of two black holes, one of them thirty-six times the mass of the sun, one of them twenty-nine solar masses, as they spiralled around each other at a significant fraction of the speed of light, before merging into a single black hole of sixty-two solar masses. The simple arithmetic delivers the energy that was expended in gravitational radiation: a lump of matter three times the mass of the sun turned to purest energy and began to storm away from the scene of the collision at the speed of light, propagating itself by distorting space-time in a distinctive pattern of waves, discernible in one orientation but not another. However colossal the distortions of space around the event itself, this pattern of violence to the fabric of space-time began to diminish as the ripples widened: that's what happens to waves when they propagate. It is why the light from the farthest stars is harder to see. It is why the propeller thresh from a distant steamer barely disturbs the fisherman's dory. But that same imagery helps me understand a little better my own fascination with physics, and the solace it delivers. Each of us occupies a small space in our own small worlds: the mathematics that underpins Einsteinian physics confirms for us that there is something beyond the horizon; that far beyond any horizon we can possibly imagine, there are happenings that create disturbances that become tremors that will one day – if we happen to be watching carefully – rock our little boat. We may not think of ourselves as in touch with the universe, but the universe keeps in touch with us. The eerie emptiness through which our small, stony planet seemingly sails blithely around its parent star, is not quite emptiness at all.

It has fabric. It moves, and we can feel it move. Or more exactly, an international community of physicists has found a way to measure a movement which can only be understood as a distortion in space-time, triggered by distant violence on a scale we can measure, but only after devising instruments of astonishing subtlety, and then spending another decade or so making them even more sensitive.

And we can do this because although waves become harder to observe, it doesn't mean they don't continue their progress. The record of the death of two black holes in one brief cataclysm rippled through space, at ever diminishing volume, for a billion years, until it reached a tiny planet orbiting a modest star in a faraway galaxy, to tell the story to a pair of detectors. Each of these detectors is made up of one laser beam, split into two and sent shooting away at right angles to each other. Each part of the divided beam hits a mirror and bounces back, to be recombined. When the two signals recombine, they cancel each other, exactly according to prediction.

There are other technological challenges to the endeavour to identify gravitational waves, to confirm the last unconfirmed predictions of Einstein's Theory of General Relativity, and to take electronic snapshots of the death of stars. We live in an active, vibrating, thumping and bumping world. Winds shake structures, earthquakes on the far side of the globe send shivers through the bedrock, sunlight makes stone and water expand: anything – the rumble of a lorry across a motorway junction, the crash of a dustbin as it hits the ground – creates movement which could be picked up by a detector so sensitive that it could register a compression measured in millionths of an atomic diameter, or thousandths of an atomic nucleus. To detect a change in space-time itself, researchers and engineers had to find ways to isolate the instruments from the humdrum hubbub of the world we experience. And they did it.

These detectors may be state of the art, and without parallel in their sensitivity, but they are based on very familiar physics: waves can reinforce each other, or cancel each other, and light – with an absolute speed, divisible by frequency or wavelength – is the most reliable measuring rod available to humankind, or to any other possible extra-terrestrial intelligence. The detectors in the states of Louisiana and Washington are called interferometers. If the reflected beams of light from an interferometer show a difference, then something must be interfering with the traffic of one of the beams. Since the laser beams are travelling through a vacuum containing only space-time, then interference is a consequence of the distortion of that space-time. If space-time was being distorted everywhere, and in every direction, we wouldn't know about it. But if the distortion came from one part of the sky, but no other, then it would be registered by one arm of the beam and not the other. Because the distortion is measured in terms of thousandths of the diameter of an atomic nucleus – which is far tinier than any atom – a measurement by one instrument alone would not of itself be very convincing.

It is worth contemplating, just for a moment, the phrase 'thousandths of the diameter of an atomic nucleus' along a laser beam that serves as a measuring rod that extends for four kilometres. This is a very small distance: how small is illustrated by scaling it up. Think instead not of four kilometres but four light years. This is the distance to our nearest star, Proxima Centauri. On such a scale, the detectors could measure the movement no greater than the thickness of a human hair. Once again, we have a demonstration of the capacity of physicists to measure the seemingly unmeasurable.

On 14 September 2015, at 09:50:45 Universal Time, a signal that became known as GW150914 was recorded at the Laser Interferometry Gravitational Wave Observatory's instrument in Livingston, Louisiana. It lasted no more than twenty thousandths of a second. The other half of LIGO, 3,000 kilometres away at Hanford, Washington, recorded an almost identical signal. There was an interval of ten milliseconds between the two signals. That is the point of two detectors. If there was only one detector the experimenters would never know for certain whether they had detected a gravitational wave, or whether a loaded truck had hit a brick wall a few miles away and set up a tremor that could be registered by the nearest detector but not the furthest. But the two detectors got the same signal in one hundredth of a second. Light travels at 300,000 kilometres a second: so what arrived at the first detector was travelling at the speed of light, and continued on at the same speed, to the other detector to tell the same story. That immediately told researchers which part of the sky the signal arrived from. And the interference pattern of light waves within that signal lasting twenty thousandths of a second matched a set of theoretical predictions that said: suppose two black holes collided? This is what the collision ought to look like, if translated into an interference pattern recorded by a laser beam.

Three minutes after the signal arrived, an automatic detector announced its importance, and the experimenters and theoreticians set to work. It took five months for the research teams to settle the matter to everybody's satisfaction. They had detected a gravity wave, and they could say with more than 90 percent confidence that it had been caused by the collision of two entities so dark that no light could ever escape from, or be reflected by, their surfaces, composed of what had once been matter forever diminishing towards infinity, but distorting space-time around themselves in ways that could be calculated. And the researchers were confident enough to give these two black holes mass: the smaller of them a black hole twenty-nine times the mass of the sun, give or take four solar masses, and the other thirty-six times the mass of the sun, give or take five solar masses. The same calculations agreed by the 1,000-plus scientists who make up the LIGO team, and their European partners, delivered a guess as to how far away the signal began. The cosmic traffic accident happened at a distance of 410 megaparsecs. A parsec is a term based on two other terms, parallax and second. This is terminology that doesn't mean much to most of us, but it helps to know that this is the distance light would travel over a period of 3.26 years. Multiply that by 410 million, with once again, a band of uncertainty: the distance could have been 570 megaparsecs, but not more, and it certainly wasn't less than 330 megaparsecs. And in this ponderous pronouncement of certainty, complete with the estimates of potential uncertainties, the LIGO announcement made history. It settled a question that had been unresolved for a hundred years, it vindicated the faith of a community of physicists who had persuaded governments to invest in ever more expensive detection systems, and it opened the way for a new kind of astronomy.

It confirmed a prediction made possible by Einstein's reasoning, but one that he, and most physicists who followed him, did not expect to see ever confirmed. By the time the signal arrived, the question was, in any case, academic: Einstein's Theory of General Relativity has been repeatedly confirmed in countless observations, using measuring technologies that have become more exquisitely precise with each decade. Just one prediction remained untested. By the time LIGO had detected signal GW150914, most physicists knew in an intellectual sense that gravity waves existed. The question was: could they reliably be detected? And the answer to that was a resounding yes, with a great deal of extra information. And humankind, imprisoned as a species on the surface of its own little planet, a species that began its inventive history with little more than a flaked stone as a hand axe, a species of limited perspective but with an uncomfortable awareness of its own mortality, could nonetheless deploy physics to reconstruct a drama that had happened long before the first glimmer of any complex life on Earth. First, the technology and the theory between them could distinguish the kind of cosmic cataclysm that could send distortions in space-time rippling across the universe: these were made by black holes, not colliding neutron stars or sudden supernovae or any other conceivable or yet-to-be conceived candidate.

Furthermore, the pattern of waves revealed by the interference signal told some of the detail of the collision: that, for instance, at the beginning of the 0.02 seconds recorded by LIGO, the two black holes were more than 1,000 kilometres apart, and orbiting around each other fifteen times a second. And of course, getting closer by the microsecond. Near the end of this ever-so-brief signal, the two holes were spinning about each other at 250 times a second and the collision took place at almost half the speed of light itself. As each black hole neared the other, it lost mass in the form of energy that sped away at light speed as evidence; the speed at which their orbits collided gave physicists an idea not just of their mass but of their dimension: the smaller had a diameter of 174 kilometres, and the larger 216 kilometres (in both cases of course such measurements are approximate). So this one tiny signal recorded by one pair of detectors delivered not just direct evidence of gravitational waves, but reliably inferred evidence of the collision of two black holes, each with a mass and dimension that could be estimated, and then on top of that, it confirmed that black holes could co-exist as binary partners, invisible entities locked in a dance that could end in some unthinkable consummation. It wouldn't be something that happened very often, or near enough and often enough to be detected every day, but it proved a point. The theory was correct and the technology worked. And, for physicists, it was not the end but the beginning.

A new form of astronomy was possible: a way of peering into the heavens using something other than light, a way of registering amazing, one-off phenomena involving forces and masses of the kind so far permitted only as things that could happen in theory. It would happen again, and again, and each event would tell a different story. LIGO detected its second gravitational wave signal in June 2016. By 14 August 2017, the day LIGO recorded event number four – a signal known as GW170814 – a third detector was ready: a collaboration known as Virgo had completed its upgrade and fine-tuning not far from Pisa in Italy, in time to register a collision between two orbiting black holes, one thirty-one times the mass of the sun, the other twenty-five solar masses. The signal was recorded in Louisiana first, and then in Washington eight milliseconds later, and then by the Virgo collaboration in Pisa in fourteen milliseconds. With three detectors on the case, the researchers were able to narrow down the area of sky from which the signal came, and identify it as even more distant than the first ever event: a distance of 550 megaparsecs or 1.8 billion light years. And this time, they could draw a conclusion about the way the waves were polarised: if black holes merge by spiralling around each other, they must do so in a two-dimensional plane, so the distortions of space-time should be more assertive along the axis of that plane, and more demure at right angles to it. But that's an assumption. With three detectors on the case, it might be possible to make better measurements of the geometry of collisions on a cosmic scale. The Japanese are now working on a fourth detector.

The European Space Agency has taken up the challenge with a project called LISA: a fleet of spacecraft operating in ultra-precise formation, measuring distortions in space-time across regions far greater than could ever be achieved on Earth. LISA stands for Laser Interferometry Space Antenna, and at the time of writing researchers have just switched off the LISA Pathfinder mission, a sixteen-month experiment in steadiness and stability to a scale beyond the wildest dreams of science fiction: the thrusters on the spacecraft could keep it on an orbit to within a predicted accuracy of ten billionths of a metre. In gravity wave astronomy, the longer the measuring rod and the surer the degree of precision, the better the information. The full LISA project would involve three spacecraft in the formation of an equilateral triangle, each 2.5 million kilometres distant from each other, each linked with the others by laser beam. If the mission goes ahead, the three spacecraft won't be launched until 2034. If it does go ahead, it will be the first telescope to observe the engine of gravity at work across the entire universe.

VI

The Planet Factory

We should pause at this point. The distortion of space-time recorded so far by gravity-wave detectors follows a pattern of big science: the financial investment is large, the detectors are colossal and the returns are vanishingly small, in the most literal sense. A distance measured in thousandths of the diameter of an atomic nucleus is not just small, or even vanishingly small. It is so small that, had it not been registered almost simultaneously by two such detectors, no one would or could have accepted it. A change in distance measured in terms of the thickness of a human hair over four light years – an analogy calculated for the benefit of the media in 2016 – is not, in any practical sense, a change at all. For physicists and engineers, the announcement was a triumph: their design, construction and confirmation technologies had done what a few decades ago must have seemed impossible. They had proved that gravity waves could be detected.

They had of course achieved other firsts: they had shown that black holes could exist as binary pairs, and they had again confirmed the predictions inherent in Einstein's

Theory of General Relativity and – a nice touch – done so during the hundredth anniversary of the first formulation of the theory. But by 2016, almost all the logical outcomes of general relativity had been confirmed, repeatedly. From one perspective, the confirmation of gravitational wave events must have seemed like a climax with its own anti-climax already incorporated, which may be why many scientists greeted the news not as the conclusion to a fifty-year-challenge but as the moment when a new era of astronomy could begin.

And it could be said to have begun not so much with the first observation, or even the ones that followed, but with something in every sense more spectacular: a collision between two neutron stars. This matters because in a black-hole collision, there is no debris: the invisible mass just becomes even more invisible. But neutron stars must be the source of some of the elements of which habitable planets are fashioned. It is a commonplace among astronomers that we are all composed of stardust. But how stars make the fabric of planets, and thus people, is a story still incompletely written. Neutron stars are the densest visible things in the universe: huge stars that have burned up their fuel and contracted under their own surviving mass to a point where they are no longer composed of atoms, as we understand atoms. Physicists used to think of the atom as a little solar system: a dense and sturdy nucleus at the heart and around it a cloud of featherlight electrons in ever higher orbits. That analogy was of limited use, and doesn't really describe the topography of an atom, but the useful thing it does is highlight the nature of an atom: it is mostly empty space, maintained by forces almost immeasurably stronger than gravity. Until, of course, you amass enough atoms in one place. Then the crushing begins. A star maybe twice the size of the sun – which has a diameter a hundred times that of Earth – is crushed to something that might measure ten kilometres across, spinning several hundred times a second. A spoonful of the stuff of this star would weigh billions of tons. If the original inflated star had been three times the mass of the sun, this shrinking would continue: the final collapse would be into a black hole. We might never know it had been there at all. Neutron stars, however, stop just short of that: light can escape from them. We know some of them as pulsars, emitting short bursts of radiation as they spin, up to 700 times a second. Then think of a pair of neutron stars waltzing around each other, in what astronomers call a binary system.

For at least 200 years astronomers have been recording binary stars, star systems that orbit each other. There must be millions of neutron stars in the Milky Way alone, and many of them must be in a binary system, and sometimes both binary stars will therefore be neutron stars. Then think of them inching, or being nudged by some other star's force, ever closer until, like romantic partners in a movie, they rush into each other's arms. In 2017, researchers around the world recorded such an embrace, and the first to do so was the LIGO–Virgo partnership. Swiftly, other observatories focussed on the patch of sky that signalled the dance of death of two neutron stars in a nameless galaxy, 150 million light years away. Within days, researchers had started to use cumbersome phrases like 'multi-messenger astronomy'.

The obliteration of the two neutron stars, probably into a black hole, was registered as a fleeting gamma-ray burst, the brightest thing in the heavens, and then in all the other wavelengths: X-rays, ultraviolet, optical, infrared, radio and so on. They also confirmed – there being enough information carried in these wavelengths of light to confirm such a belief – something astrophysicists already knew, but had yet to observe directly: that immensely energetic big bangs and bursts in immensely heavy stars constitute the machinery that create the heaviest elements. We and everything we see are so much stardust.

The universe began with an endowment at birth of hydrogen and helium. All the other elements were forged in the thermonuclear furnace of a star, which once dead had to explode, or smash into another star, to release its nitrogen, oxygen, carbon and so on. But ordinary stars stop the process at iron. Stars the size of the sun burn out before they can go beyond element number twenty-six in the periodic table. Iron is the commonest element on Earth – the planet has a core of iron – but the theory of the thermonuclear fusion process that drives main-sequence stars like the sun cannot explain the manufacture of the heavier elements. The theorists had to stop at a kind of stellar iron gate. The heavy-duty elements needed some other explanation, and so supernovae and other superlatively destructive events had to be invoked to account for the creation of all the other and often rarer elements up to uranium. In the crash-bang-wallop of this observed collision of two neutron stars, immediately dubbed a kilonova, astronomers and spectroscopists recorded the creation of heavy metals on an epic scale. The shrapnel blown away from the heart of the collision – away from the black hole freshly formed by the collision – included a mass of platinum, gold, uranium and so on to the mass of about 16,000 Earths. One astronomer boldly hazarded that the collision may have created the weight in gold equivalent to a hundred Earths.

In a cheerful demonstration of the narrowing gap between fairy-tale fun and scientific observation, in 2012 exoplanet hunters at Yale University identified a planet called 55 Cancri e, much bigger than Earth, orbiting a star called 55 Cancri, about fifty light years distant. It seemed to be composed mostly of diamond and graphite.

Forbes magazine set out to value the planet at $26.9 nonillion. That's a figure with thirty zeroes in it. The valuation does not include the cost of bringing such a planet to market, nor the response of the diamond market to a shift in the demand / supply ratio on such a scale. But that is not the point. The point is that because of the study of an event in a distant galaxy, we now know a lot more about the fine detail of the fabric of planet Earth.

We owe everything to a traffic collision involving at least one superstar. A neutron star collision in a galaxy like the Milky Way might happen only once in 10,000 years. In an older galaxy, such collisions might happen ten times more often. In an observable universe that might be home to 100 billion galaxies, neutron star collisions should be an everyday event, but even instruments as sensitive as LIGO would be able to record only the nearer such happenings. The observation showed that, in every sense, the physicists had struck gold. And yet, of course, they had expected to do so. Theories said an instrument like LIGO should be able to record invisible collisions between black holes and smashes between neutron stars brighter than 1,000 suns. But the instruments' backers and supporters and builders had something much more ambitious in mind. 'For me the most exciting thing is we will literally be able to see the Big Bang. Using electromagnetic waves we cannot see further back than 400,000 years after the Big Bang. The early universe was opaque to light. It is not opaque to gravitational waves. It is completely transparent,' one of them told me at the time.

> So literally, by gathering gravitational waves we will be able to see exactly what happened at the initial singularity. The most weird and wonderful prediction of Einstein's theory was that everything came out of a single event: the Big Bang singularity. And we will be able to see what happened.

This sounds like one of those things scientists tend to say to celebrate some landmark achievement: understatement in such circumstances has never been a good idea. But I think he meant what he said, just not in any obvious way. We have been here before. Everything we see is old news: the face in the mirror is the face that existed a few billionths of a second before you look. The logic of ever better and bigger telescopes is that that they will see further, and therefore further back in time. Go back far enough and at some point nearer the beginning, say the theorists, the universe goes dark. The astronomers' own name for this moment is 'the Dark Ages'. And long before the instruments in Louisiana and Washington and northern Italy, physicists had proposed that gravitational wave detectors might reveal details about the early history of the universe in the long, dark years before galaxies formed, and light began to shine across the heavens. But that's a satisfaction to come. What becomes clearer from the expression of such hopes is: what we do not know.

We do not know how big this universe is, or what shape it is. Things are as they are because the forces that order our lives – the so-called laws of gravity and thermodynamics, the electromagnetic wave equations, the ratios between the various forces that shape the universe and so on – are as they are now, and presumably have always been so, as far as we can tell. A new orthodoxy places Earth and the solar system at the centre of an observable universe that extends as a sphere across a diameter of 93 billion light years, or 10^{27} metres or a trillion trillion kilometres. That kind of imagery ought to be helpful – it delivers a map of where we are in the heavens – but any such comfort, any feeling that at least we now know our place in the great scheme of things, vanishes when we think about it. It starts with the word 'observable'. This is a recognition that there is a limit to how far we can see, and beyond it there is more, yet to be seen, and beyond that, still more that may never be seen. We are the centre of creation only because we can see in every direction, so it appears as if we are in the centre. Some notional sentient observers on a habitable planet in a galaxy far beyond our own horizon may look out beyond their own orbiting lump of mineral, water and gas and see much the same vast spacescape extending for much the same distances. The logic of such thinking is that some of what these observers see would overlap with the regions we can see, and some would not. If our two horizons intersect, and they see what we see, then the forces that shape our observable universe are the same for them.

It's hard to see how this could not be the case. Cosmologists have been playing with the notion of different values for the electromagnetic, gravitational and other forces that shape us for many decades. In one thought-provoking account, called *Just Six Numbers* (1999), Martin Rees identifies the mathematical relationships that make this universe stable. For instance, if the ratio between the strength of the electrical forces that hold atoms together and that of the force of gravity that holds the universe together was even slightly different, we would not be here. One of his six numbers is a very simple one: D = 3, where D stands for dimension. That is: we can look vertically, across and along. There are three directions in which we can travel. Those are our three dimensions. Life as we know it would not be possible if that number were two, or four. But mathematical physics has no problem with a higher number of dimensions: in fact, some of the reasoning about why the universe is as it is, proposes that at some level, higher dimensions must exist. Logically, there is no reason to expect that the conditions that apply in our part of the universe, and the forces that created those conditions, are universal in the sense that we casually use the term.

If our observable universe began with a tiny random event involving a virtual particle and a false vacuum followed by a very brief period of cosmic inflation, why would such a thing happen only once? Why would it not happen somewhere else, and at some other time (if words such as 'time' and 'somewhere else' have any meaning in this context)? Is there some other part of this universe in which cosmic inflation is still happening? Or are there other universes bubbling off this one? In the creation we see around us, anything that happens once can happen again. Does that apply to the whole bag of tricks? Is there an eternity of universal fabrication, a kind of foaming fountain of universes, each with its unique laws of physical order, set so that some of these wink into existence, and then collapse before their substance can set into galaxies and stars and planets and people, while some endure a little longer, to provide a home for entities we can only describe as unknowable, even if somehow we can imagine them? Do these universes keep their distance from us? Or are they contiguous, adjacent, at right angles to or in some way connected with us? Are there other dimensions? Other worlds we might one day begin to know and yet other worlds we cannot know, all somehow in our vicinity?

In 1895 H.G. Wells achieved enduring fame when he wrote his novella *The Time Machine*, yet few know his very different little novel of the same year called *The Wonderful Visit*. Wells' protagonist is one of those amiable sportsman-naturalist vicars that seem to have characterised eighteenth- and nineteenth-century England. He hears of a strange and beautiful bird spied flying over the moors, which he suspects is probably a flamingo. Then one day out walking, he sees something colourful fluttering in front of him and – since he is out with a loaded gun – fires 'out of pure surprise and habit'. He brings down 'a youth with an extremely beautiful face, clad in a robe of saffron and with iridescent wings, across whose pinions great waves of colour, flushes of purple and crimson, golden green and intense blue, pursued one another as he writhed in his agony'. The vicar had shot and winged an angel, an angel as defined by a tradition of religious art that extends from the Renaissance to the Pre-Raphaelites, and being a humane naturalist vicar he does the decent thing, and carries him off to the vicarage until the injury to the pinions has healed. The story is a social comedy, and the comedy rests on the idea of interlocking universes, one in which we imagine entirely spiritual beings called angels and then imagine them with wings and long robes, and another in which a wounded angel in a state of severe shock is nevertheless able to say, in English 'A man! A man in the maddest black clothes and without a feather upon him. Then I was not deceived. I am indeed in the Land of Dreams!'

The story, slight as it is, touches on all the religious bigotry, prurient suspicion, social anxiety and fear of socialist revolution you might expect to observe if you introduce a beautiful youth in a ravishing frock to a small West Country parish in the last years of Queen Victoria. But in the course of this comedy Mr Angel and the vicar of Siddermorton get to discuss their intersecting worlds, and remind us that the idea of coincident universes, perhaps dimly perceived only as dreams, is not a new one:

> 'It is confusing,' said the vicar. 'It almost makes one think there may be (ahem) Four Dimensions after all. In which case, of course,' he went on hurriedly for he loved geometrical speculations and took a certain pride in his knowledge of them – 'there may be any number of three-dimensional universes packed side by side, and all dimly dreaming of one another. There may be world upon world, universe upon universe. It's perfectly possible. There's nothing so incredible as the absolutely possible. But I wonder how you came to fall out of your world into mine ...'

Accordingly, a trio of physicists proposed in *Scientific American* in 2014 that the world we now occupy – or think we occupy – is in fact a kind of hologram. Our three-dimensional universe is, they write, 'merely the shadow of a world with *four* spatial dimensions'. In their scenario our entire universe is an offshoot, a product or a blast from a stellar implosion in a suprauniverse that 'created a three-dimensional shell around a four-dimensional black hole. Our universe is that shell.' It sounds as though Wells might have got there first. As Sherlock Holmes once observed in *The Sign of the Four*: 'How often have I said to you that when you have eliminated the impossible, whatever remains, however improbable, must be the truth?'

Cosmologists have proceeded thus far by working out what, mathematically at least, is impossible, and in a universe as vast and strange as this one, who is to say what is improbable? We have contemplated the birth of the cosmos from a lump of quantum foam, or a virtual particle trapped in a false vacuum, and proposed cosmic inflation as the logical explanation behind the apparent sameness of the universe in every direction – why would anything be improbable?

In 2015, a physicist proposed that the fabric of the universe – that is, the fabric in which galaxies, black holes and stars and planets are embedded – could be considered as a fluid. Fluids have viscosity. Water isn't very viscous, treacle is horribly viscous. Space-time is probably as near to not viscous as a fluid can be and still flow. Space expands.

That has been an observed phenomenon for almost a century. Suppose you imagined a universe in which space started to expand at an ever faster rate? How fast could this get? In 1998 physicists observed accelerated expansion: they were using stars of a characteristic brightness called type 1A supernovae as a standard candle. That is, the fainter these were, the further away they would be. With them, you can take the measure of the cosmos. In 1998, researchers decided that something was wrong with their picture of how things should be: the farthest standard candles were receding at a rate beyond all their predictions. And after a bit of thinking, they delivered a new idea. A force, a kind of anti-gravity – a quintessential property that lay in space-time itself – had taken over, and was driving an ever-accelerating expansion of the universe, a continuous acceleration that we could not detect in the immediate universe around our galaxy and its companions, or the supercluster of galaxies in our sector of the universe, but became inescapably obvious at its far horizons. Within months, the cosmologists had derived an unidentifiable force called dark energy, inherent in space itself, to accompany the dark matter they already knew about, but of course, could not in any useful sense, know. Which once again, brings us back to the properties of the cosmos itself: does it have viscosity? If so, and if dark energy goes on driving the expansion of space at ever greater speeds, what happens to the tenuous and (remember this) hypothetical viscosity? Comfortingly, we will be put out of our misery. The universe in this scenario will not expand to infinite speeds to become so tenuous in 100 trillion years that we will be lost in utter cold and darkness. Rather, it will proceed to a certain level of pressure and density and start to rip itself apart. There is a limit beyond which cosmic viscosity cannot sustain itself, and at that point, matter, time and space snap like a stretched rubber band. Earth explodes. So does the solar system, and the galaxy, and all the other galaxies. The universe will have reached what the theorists call 'an extreme end state'. This could happen about 22 billion years from now. Since, of course, long before then the sun will have flared up and incinerated all the nearest planets, we have nothing to fear. We will all be dead anyway. But we are in the happy position of being able to foretell the spectacle at the end of the universe. We will all go out, like a light. It will be like the close of Arthur C. Clarke's 1953 short story 'The Nine Billion Names of God', the one that ends just as the high-speed computer installed in a Tibetan lamasery calculates the 9 billionth name of God, to be pasted into the inventory of the names of God, and the two Westerners who installed the instrument slip away, congratulating themselves on having left before the monks discover that the computer program has failed them, and that the universe will not end with the 9 billionth name, only to look up and see that 'overhead, without any fuss, the stars were going out'.

I used the word 'happy'. Boethius used the word 'consolation'. It isn't easy to understand quite why speculation on this scale delivers pleasure, except, once again, to return to human preoccupations so profound and enduring that we have framed them in religion, and in philosophy, in art and literature and science and of course in science fiction, and they remain roughly as framed by the Bronze Age poets who first began to record, in written and enduring form, the mystery and wonder of creation. Both in religion and in science, we play with possibilities and compose stories to explain what might be. The pleasure of what could be called untestable cosmology – the cosmology of the multiverse, of string theory, of cosmic inflation, of singularities, of multidimensional universes – is not really separable from some kinds of theological speculation.

Cosmologists have proposed the ideas of dark energy or cosmic inflation because they address a logical but intractable problem and offer a possible solution. They deliver a way of tidying up a story, entirely at a provisional level: these things might be true and if so it follows that we can therefore explain this or that observed state on the basis of such assumption. Christian theologians recognised a problem long ago of the fate of those who lived and died before they could be redeemed by the sacrifice of Jesus Christ, or those who sinned but tried very hard not to sin again. They proposed supernatural geographies that include limbo and purgatory as a solution. The former would be a place of loss – but not suffering – for those immortal souls who had been virtuous in life but died before they could have ever known the truth of Christianity, and Dante placed Virgil in this earthbound limbo, to serve as a guide to hell and purgatory, where of course the souls of those who were not saints could be purged of sin before entry into heaven. Neither limbo nor purgatory are described in any Holy Scripture, but nothing in the Old or the New Testament forbids them. That is, they represent ideas that cannot be tested by living humans in the present state of knowledge, but useful ideas for those who want to think about the religion they profess.

And that does indeed fairly describe some instances of cosmological thinking, with this difference: that the whole point of scientific hypothesis is to test it to demolition. If the hypothesis resists demolition, and goes on providing possible explanations for things that are observed, then it becomes an increasingly useful way of thinking about the world. That doesn't mean that it is surely the right way, or the only way to think about the world. The difference is that religious devotion demands unquestioning faith. The practice of science demands a state of mind that is always open to doubt. Those of us who are neither scientists nor particularly religious can enjoy both, but it may be easier to enjoy science because at some level, it represents truth in a way that can be, or one day could be, tested. Instead of dogmatic answers, it delivers provisional responses. Dogma admits no equivocation: you must either take it or leave it. There can be no questions, which is an impossible stipulation for the questioning mind.

The Consolation of Philosophy is a text by an early Christian that continues, right to the end, to question the difference between the invocation 'God knows' and the question 'Yes, but how do I know?' In its art, it manages beautifully to address the idea of an eternity in which all successive human actions and choices can be known and observed simultaneously, but which still commits humankind to responsibility for the choices it makes and the acts it chooses to perform. What will happen is not the same as divine will. This extended reverie in the face of imprisonment and impending death does not represent scientific thinking, but its approach to reasoning is not separable from the reasoning that supports science. And it is a reminder – as if we needed reminding – that the so-called, and once frequently invoked, gap between art and science might be a misunderstanding, or perhaps not a gap at all. Which, oddly, brings us back to Copernicus.

Even if we *know* that we are not the centre of the universe, we *think* as if we are: as citizens of a single planet, we have no other perspective. We can see only so far in every direction, therefore we must be at the centre of the world we inhabit. That is a way of saying our perspective is necessarily subjective. But there is another subtlety. We think we see the 'real' world. In fact, our eyes do not do the seeing. We have already been here: the brain delivers the vision, registers the imagery, and constructs the picture of the world for us. We do not see, instead we think we see. Since we are all different, that means that what each of us sees, or hears, or smells, tastes or touches is unique. There is no way in which I can confirm that the red that you say you see is the same hue as the scarlet that I think I see. We all of us live in a set of unique and uniquely egocentric worlds: we have no other choice. You can buy and play with virtual-reality headsets that will deliver a seemingly shared experience, but we tend to forget that at birth each of us was born with a pre-installed, shrink-wrapped, bone-protected virtual-reality headset, in which we project for ourselves our version of reality, and no other. We act as if our experiences are shared, but ultimately each of us remains a single viewer in a multiscreen auditorium, trying to take in the entire multiple experience and concentrate while doing so on this or that episode of the venal soap opera, political drama or social comedy playing before our eyes. We believe we stand, perhaps in a gallery devoted to art, before a painting by Henri Matisse: the evidence that we do so, however sensual, is processed and composed and experienced in this strange mass of nervous tissue. I believe I stand in a French provincial museum, looking at something by Matisse: perhaps I only think I do. And art is the best medium for these shared experiences: novels, paintings, films, plays, radio comedies, concerts, operas, movies, rock songs, ballads, advertising hoardings, adventure stories, biographies, histories and television serials deliver, around the clock and every day, the experiences and perceptions of others, and each is a reminder that although we are all unique, we can all in some satisfactory way attempt to understand how other people think. Subjective we may all be, but we understand the subjectivity of others.

At its most helpful, science does the reverse: it delivers or at least aims to deliver an objective assessment of the planet on which we live, and the other lives with which we share this space. It delivers names for things, and histories, too. Best of all, it delivers definitions on which humans can gratefully agree, and defines properties that we can exploit and in every sense enjoy. Let us take a down-to-earth example.

Limestone is calcium carbonate; so is chalk. Both are sedimentary rocks, and they differ only in their degrees of hardness. Limestone baked to release carbon dioxide becomes a calcium oxide that is the basis for cement or for the quicklime sometimes used in steel making and other processes. On its own, limestone can be used as a building material. It must not be confused with gypsum, which is calcium sulphate, which after treatment can be used to plaster walls that have been cut from limestone blocks and secured with a cement-based mortar. Quicklime burns with a brilliant white glow, and was used to illuminate Victorian music halls and theatres, which is the origin of the term limelight. Limestone provided the stones and the mortar for the Great Pyramid of Giza in Egypt; it provided the building materials for the great cathedrals of Christian Europe and the castles of medieval Britain, and it shored up the economy of Victorian England. In 1865, in the novel *Our Mutual Friend*, Charles Dickens introduces two clandestine witnesses – the solicitor Mortimer Lightwood and the barrister Eugene Wrayburn – to the Limehouse public house known as the Six Jolly Fellowship-Porters, and their police inspector companion tells them for a cover story that '"You can't do better than be interested in some lime works anywhere down about Northfleet, and doubtful whether some of your lime doesn't get into bad company as it comes up in barges."' And Lightwood reponds,

> 'You hear, Eugene?' said Lightwood, over his shoulder. 'You are deeply interested in lime.'
> 'Without lime,' returned that unmoved barrister-at-law, 'my existence would be unilluminated by a ray of hope.'

At a number of levels, science illuminates art: we cannot understand our history or our literature without a shared understanding of the fabric from which our shared world is composed. And the reverse is certainly true: we tend to understand and remember and value science in ways that do not meet any popular definition of the scientific method. I cannot think of limestone just as a sedimentary rock that contains more than 50 percent calcium carbonate, relatively soluble in meteoric water: even that simple statement tells a 100-million-year story of hot sunlit oceans long ago, oceans rich in coccolithophores, tiny creatures with calcite shells that after death or digestion by predator fell in a steady drizzle to become chalk, or make up the bulk of a limey mud, into which might be preserved the whole skeletons of marine reptiles, in their turn to be covered by precipitates from the ancient oceans in the form of chemical rather than biological calcite: a much, much warmer world, in which land-based dinosaurs stalked the woodlands of what would now be Alaska or Antarctica. In 1895, H.G. Wells ended *The Time Machine* by wondering where his Time Traveller had got to:

> Will he ever return? It may be that he swept back into the past, and fell among the blood-drinking, hairy savages of the Age of Unpolished Stone; into the abysses of the Cretaceous Sea; or among the grotesque saurians,

> the huge reptilian brutes of the Jurassic times. He may even now – if I may use the phrase – be wandering on some plesiosaurus-haunted Oolitic coral reef, or beside the lonely saline lakes of the Triassic Age. Or did he go forward, into one of the nearer ages, in which men are still men, but with the riddles of our own time answered and its wearisome problems solved?

Oolite is a carbonate rock composed of shell fragments. The only reason we know that there were such things as plesiosaurs is because they still haunt ancient reefs: their bones have become part of the seabed, 100 million years later to be exposed as a chalk or limestone cliff face. We see them only because millions of years of rain and surface have dissolved or swept away the softer limestone, to expose the ghostly structures of ancient sea monsters. Wells enlisted his plesiosaurs in the service of art. We all instinctively see science – once we understand the story it tells – as part of some great, extended epic poem, or a sublime structure; a celebration both of the entire temple of creation and separately of the beauty and fascination of every pillar, arch, niche, cornice, buttress, paving, architrave, capital, caryatid, baluster, frieze and arcade that makes up the whole. And it is important to recognise that we marvel at both the whole and the parts for their beauty, and for the thrilling, dramatic stories they tell.

Right now, as I write this in an almost perfect English autumn, the most conspicuous feature through the nearest window has been the sudden appearance of spiders' webs: they sit, on bushes of rosemary or spurs of climbing rose, strung with tiny drops of morning dew like jewels on gossamer. Does it make it any less magical a sight to know that this web was spun from a protein inside the spider? And that this liquid silk has a most unusual property; that the instant this liquid is exuded and tautened the molecules snap into formation to make the strands of the web? And that weight for weight, and diameter for diameter, this silk is stronger than steel, or harder to break than rubber? And that a tiny garden spider can spin 200 feet of the stuff, in up to six grades of thickness, with more than 600 attachments in the web, all in less than an hour? Does that make a web aesthetically less marvellous?

Not long ago, it was possible – and it may still be possible – to hear from people who say that science is soulless. To which I can only say, somewhat mildly, it doesn't seem like that to me. The more you learn about anything, the more mysterious and marvellous it becomes. The paradox of big science – the science that tries to take the measure of, and transcribe the history of, the universe itself – is that the more intractable the mystery, the more pleasure we get from the little things we can learn of it, and even those tiny snatches of understanding tend to highlight the vastness of what we still do not know. And, as I have said from the beginning, humans, funded by national governments, have embarked on these missions of discovery, in the knowledge that there can be no obvious profit, no bullion to bring home from the voyage. Projects such as CERN, LIGO and many others, have a long history. Captain Cook set sail in HMS *Endeavour* to observe the Transit of Venus, one of several transnational projects to make accurate measurements of the passage of a planet across the face of the sun, and use the observations to arrive at the correct value for an astronomical unit; that is, the distance from Earth to the sun. His admiralty orders did invite him, once the Venus mission had been completed, to take a closer look at the southern Pacific, and as a consequence, Britain ended up claiming both Australia and New Zealand, and a number of Pacific atolls and seamounts. But the voyage began ostensibly in pursuit of knowledge of scientific value, and with no overt intention of gain. Nor was Britain immediately the richer at its end and there was, if anything, a loss: for the indigenous peoples of Australia and the Pacific. The Large Hadron Collider, the Laser Interferometry Gravitational Wave Observatory and the *Voyager* mission are epic acts of co-operative discovery, for which nobody, anywhere, is overtly the poorer. The gain may be intangible and uncertain, but it is available to all, and enriches us all.

Forty Years On

The *Voyager* mission has now been running for more than forty years. Its place in history was established within the first five years or so. *Voyager 2* was the first spacecraft to fly past Jupiter, Saturn, Uranus and Neptune. Between them the two missions identified three new moons at Jupiter, and *Voyager 2* went on to identify four new moons at Saturn, eleven satellites orbiting Uranus, and six new moons around Neptune. *Voyager 1* was the first to observe active volcanoes on Jupiter's moon, Io, the first to observe lightning on a planet other than Earth (again, on Jupiter) and the first to identify a nitrogen atmosphere on Saturn's moon Titan. Both spacecraft observed evidence on Jupiter's moon Europa that a liquid ocean might be sloshing beneath an outer crust of ice. If there is liquid water underneath a casing of frozen water, then something must be keeping the water liquid: a source of heat. Heat is energy. The first requisites for life as we know it – what a cliché that phrase has become, but how else could we say this? – seem to be liquid water and a source of heat. We know from amazing discoveries in the darkness at the bottom of the oceans that there is a separate world of creatures that depend not on the energy of sunlight, but from the heat delivered by brines that bubble from Earth's crust, so the dream endures: there may indeed be life in the solar system beyond Earth, and if life can evolve separately from Earth, within this solar system, then it can evolve on and colonise any planet that offers a stable supply of energy and liquid water.

The *Voyager* mission, complete with its golden record, is a reminder of a lingering faith in what in earlier centuries was called the Plurality of Worlds. If (reasoned several generations of devout but inquisitive scientific thinkers) the Creator had intended humans to live everywhere on Earth, then perhaps He intended humans or human-like creatures or resourceful and intelligent entities to live on the other spheres visible from Earth, and furthermore on all those thousands of planets that could be imagined circling other stars faraway. In 1837, taken with this idea, Thomas Dick, author of a book called *The Christian Philosopher*, composed a text called *Celestial Scenery, or The Wonders of the Planetary System Displayed, Illustrating the Perfections of Deity and a Plurality of Worlds*, which summarised the information astronomers and physicists had by then been able to assemble about the other planets. These scientists included both William and John Herschel, two great astronomers: Dick's sources were not to be dismissed. Saturn, he proposed, had a mean density half that of water. 'The globe of Saturn, were it placed in an immense ocean, would swim on the surface as a piece of cork or light wood swims in a basin of water.' But the surface, he argued, might be as dense as the rocks of Earth's crust. The globe could be hollow or filled with 'some elastic fluid', but the surface could be home to living beings. He used the available data to calculate the surface area of Saturn, then estimated that if its citizens were like those of England, there would be 280 of them to the square mile. Which gave Saturn itself living-space for a population of 5,488,000,000,000, which would be 6,866 times the number of people then on Earth. He worked out the surface area of the rings, too. 'Were we to suppose these rings inhabited (which is not at all improbable), they could accommodate a population, according to the rate formerly stated, of 8,078,102,266,080 or more than *eight billions,* which is equal to more than *ten thousand times* the present population of our globe.' The italics are his. He uses billions in the now discarded British sense of 1 million million, these days termed a trillion. The plurality of worlds continued as a debating theme. So crowded and vital is life on Earth, wrote an unidentified critic in the *Edinburgh Review* of 1855, that it seemed that:

> the Great Designer of Nature's scheme had so manifestly willed that the portion of the material universe within the scope of human observation should be teeming with living things, that it is improbable in the extreme that the same Designer should have left blank and desolate the other wide regions of substantial capacity, which are equally fitted to be the seat of similar developments, which are unquestionably kindred parts of one physical connected system, and which in extent transcend the terrestrial surface as millions upon millions in untold immensity transcend a unit.

By 1882, if Samuel Kinns – the author of *Moses and Geology* – is a guide, it was quite likely that later Victorians understood that much of the solar system was blank and desolate. 'The old books on astronomy speak in glowing terms of the Saturnian sky as seen by its inhabitants, with the luminous rings and the eight satellites all shining at the same time; but this, though very poetic, must now be given up as contrary to scientific fact, since Saturn is not yet in a condition suitable for habitation,' wrote Kinns. 'On the other hand, it may be the centre of a miniature solar system, and its attendant worlds may be adorned with every natural beauty to enchant their inhabitants.' I quote these changes in nineteenth-century speculation as a reminder that whatever we think now about the condition and structure of the universe – and its component parts, including our own solar system – those who live a century or so from now will not see things as we do. Kinns contemplates the little 'system of worlds' that is Saturn, its rings and satellites 'travelling around the Sun at the rate of 514,000 miles a day, and yet each one retaining its appointed place and performing its appointed motions', and says we should marvel at the power of the Creator who could ordain laws and control this 'prince of planets' around one orbit of more than 5 billion miles in fewer than thirty years, or more correctly 10,795 days, five hours, sixteen minutes and five seconds 'with such precision that we can fix its arrival at any particular point'.

And of course, he was right: NASA, through its Jet Propulsion Laboratory in Pasadena, and using equations devised by Isaac Newton, and observations by astronomers from the previous three centuries, timed the arrival of Saturn and the arrival of *Voyager 2* at the same point in the solar plane with what now looks like exquisite precision. Many of us do not now marvel at a Creator who made these laws: if anything, we wonder at our ability to understand that there are laws, and that we can interpret them and apply them. 'The most incomprehensible thing about the universe,' Albert Einstein said, 'is that it is comprehensible.' The encounter itself would have been a moment to celebrate. *Voyager 1* flew within 3,800 kilometres of Saturn's moon Titan on 11 November 1980, and a day later sped 64,200 kilometres above Saturn's clouds and took 16,000 photographs before beginning its journey out of the solar system. *Voyager 2* made its closest approach to Saturn on 25 August 1981 and by then, the world was watching: radio and television broadcasters had driven their studio vans to the Jet Propulsion Laboratory; science writers and broadcasters arrived from all over the Americas, and from London, Paris and Tokyo.

Voyager, and planetary research, had just started to become an intellectual adventure for the wider world. *Voyager* confirmed conjectures by William and John Herschel that the thickness of Saturn's rings could not be more than about 250 kilometres; it confirmed suggestions by nineteenth-century mathematician James Clerk Maxwell that the rings were composed of 'an indefinite number of unconnected particles'; and it supported proposals by later astronomers that these lumps of dust and water ice were in some way 'shepherded' by smaller moons. Then it continued its journey into the unknown.

And so did we. The journey that followed could be characterised as small and hard-won advances in certainty, some of them delivered by *Voyager* and other missions, accompanied by an ever-widening appreciation of the things we did not know, and perhaps could not even have imagined not knowing. Within two years, the information absorbed by the Saturn flyby had triggered questions so compelling that the European Space Agency and NASA began to discuss a joint mission to Saturn and Titan: it became *Cassini–Huygens*, which arrived at Saturn in 2004. The lander *Huygens* touched down on the frozen methane surface of Titan in January 2005, to fall silent within seconds. *Cassini* surveyed the Saturnian system for thirteen years, before plunging into Saturn itself in September 2017.

While the solar system became the focus of one branch of physics, another community began to feel more confident about the wider cosmos. By the time *Voyager 2* flew over Uranus in January 1986 to record eleven new moons, physicists had begun to talk confidently about a 'theory of everything'. That did not or could not mean that they expected ever to know everything, but they did expect, within a decade or more, to have understood the big picture: how a universe is born, why its history took the path it did, why it seemed hospitable to life, how and perhaps when it might end. In 1988 Stephen Hawking published *A Brief History of Time*, a book that stayed on the bestseller lists for more than 270 weeks, and was translated into forty languages. It promised, in its closing lines, that were a complete theory ever realised, we would know why we and the universe existed. 'It would be the ultimate triumph of human reason – for then we should know the mind of God.' He was not, at the time, the only physicist or astrophysicist who thought such things. For a few years, it even looked as though he and his colleagues might have been right. *Voyager 2* flew over the surface of Neptune on 25 August 1989, to record six new moons, and then continued out of the solar system altogether, leaving only Pluto – now not considered a major planet – for future exploration. In that year, NASA scientists and engineers launched a detector called the Cosmic Background Explorer, or COBE. The name alone tells the story: this was a satellite built to monitor the cosmic background radiation left behind by the Big Bang, the oldest light in the universe. This is, effectively, the temperature of intergalactic space, and by 1992, mission scientists had identified exquisitely small variations in this cosmic temperature, unevenness on a scale that could explain why matter clumped into stars and galaxies, with pockets of emptiness in between, instead of extending smoothly over the entire cosmos. The spacecraft had in effect snapped a picture of the universe in embryo. The findings were also greeted as observational support for the theory of cosmic inflation. 'If you are religious, it's like seeing God,' one of them remarked incautiously at the press conference, intending only to emphasise the importance of the observations.

Once beyond Neptune, *Voyager 2*'s cameras were turned off to save power: from then on, the other sensors would tell the story. *Voyager 1* took its last pictures in 1990, including the one that would become famous as the Pale Blue Dot. In 1995, Swiss astronomers announced the first detection of a planet around another star, and introduced the word exoplanet into the lexicon; these are now being recorded in their thousands. Before the century's end, people had started to wonder if the elusive theory of everything would ever be achieved. In 1998, astronomers used observations of type 1A supernovae to add a new component to universe: dark energy. It is important to say this again: the term is only shorthand. Nobody knows what this stuff is and not everybody is convinced it exists at all, and if it is the energy of space itself, it is so small as to be beyond detection on the scale of a solar system or a galaxy, or even a cluster of galaxies. It seems to manifest itself only on a cosmic scale. If dark energy really exists, then combined with cold dark matter – that other, still-unidentified cosmic ingredient of galactic infrastructure – it accounts for about 96 percent of everything in the universe. To put it another way, all the stuff we think of as creation, the stuff we can define and measure and record – the photons and particles and atoms and molecules and asteroids, planets, stars and galaxies – in total adds up to only 4 percent of everything there is to know. The rest is unknown.

The universe suddenly started to look bigger than ever: so big as to make it possible to speculate on notions of a multiverse, or parallel worlds, or a universe infinite in size, in which everything that could happen certainly would, so that there might even be an infinity of sun-like stars, orbited by nine planets remarkably like our own. If so an infinite number of them would be populated by people uncannily like us, some of whom would bear our names and even be reading books like this one. Intelligent beings on such planets may even have launched their own spacecraft, heading out of their own star systems. We will never know. Even if such parallel worlds exist, *Voyager* will never get to them. The distances would be so enormous that words like 'astronomical' won't suffice, and anyway the two *Voyager* spacecraft are for the moment enfolded by the gravitational forces of this galaxy, and when you are wandering across the space between the stars at seventeen kilometres a second, in a galaxy that is home to 100 billion stars or more, you could be said to be not going anywhere in particular. Some of their instruments have been turned off to conserve power for a few more years.

Researchers hope to get more information about conditions beyond the solar system, always supposing you can define an edge to a solar system that you could go beyond. In 2012, particle physicists at CERN celebrated the discovery of the Higgs boson, and in that same year *Voyager 1* was declared the first man-made object in interstellar space. Both announcements were unequivocal, but both were difficult to define with any precision. *Voyager 1* was said to have passed the heliopause, which is a nebulous frontier where the stuff blasted from the sun is matched by material ejected from other stars. But each spacecraft has a long way to go before it passes the Oort cloud, which might also be regarded as another boundary of sorts. This is the supposed zone that is home to the comets that occasionally fall towards the sun, and this too could be considered a final frontier. The two spacecrafts' energy sources are fading, or nearing exhaustion. Each has radioisotope thermoelectric generators that exploit the heat of radioactive decay to deliver electrical energy, but once the isotopes have decayed, that will be the end. Each is equipped with hydrazine thrusters, little tanks of gas squirted from nozzles that – Newton's laws at work again – mean that each can change course, or turn its antenna or shift the angle of a camera. That is why, to be sure of extending the life of *Voyager 1* even a little, NASA engineers in 2017 tested a set of spare thrusters on *Voyager 1* that had last been used in 1980. If they need to change the spacecraft's orientation, to look at some unexpected phenomenon far from the last known orbiting objects in the solar system, they now have another way of doing it. But at some time after 2020, even questions of encounters with the utterly unexpected will be academic. *Voyager 1* and *2* will be silent. They will continue to speed through the galaxy, mute and sightless, each containing an identical little gold-plated statement about planet Earth in 1977. Like LIGO and the Large Hadron Collider, and the Hubble Telescope, and almost any major scientific project you might name, they represent communal attempts to resolve specific puzzles, each of which is part of a set of much larger questions: where did the universe come from? Where did life come from? Where did we come from?

The first of these may never be answered, if only because it may be an illogical question anyway. We are imprisoned within the universe, a small and very late component fabricated by physical forces we understand only to a limited extent. To know where the universe 'came from' would be, in every sense, outside our capacity. But at least we can begin to frame the questions, and propose hypothetical answers. The second is not likely to be answered simply by a study of life on Earth: the creature we now think of as the Last Universal Common Ancestor came to life, stomped around, and destroyed any evidence of any precursors or failed earlier attempts. But the universe seems to be rich in organic chemistry. A little European Space Agency lander called *Philae* touched down in 2014 on the surface of a comet, to identify sixteen organic compounds. A meteorite that fell to earth in Australia in 1969 turned out to contain around fifteen amino acids: these are the building blocks of proteins from which flesh and blood are fashioned. How such things might have assembled themselves into the first bacterium remains a profound mystery. Were we to witness such a process on another planet, we might be able to discern some of the mechanisms that somehow confect energy-consuming, ever-reproducing entities from accidental biochemistry, and perhaps begin to understand a bit more about the laws that govern life, if there are any, and if there is life anywhere else in the universe. The third question ought to be answerable, in the sense that we can observe a prehuman history, and a family tree of extinct hominids, some of them with behaviour that we recognise as not unlike our own. But of course, we may destroy ourselves long before we get to answer that question too.

The death of the sun will be the death not just of us, but of everything we learned and everything we tried to do: our supercolliders and gravitational wave observatories will be dismantled into their constituent atoms, along with our libraries, our museums, our sarcophagi and our arsenals. If we are alone in the universe then we will have vanished and a heedless and uncaring universe will never even have known we were here.

I end as I began: long after we have slipped away, and the death throes of our nearest star have erased any evidence that we ever existed, *Voyager 1* and 2 will still be sailing through the vastness, blind, mute and yet with a story to tell, travelling in hope rather than expectation. Each of them carries the same message: the words are attributed to President Jimmy Carter, but they'll do very nicely for me, too:

> This is a present from a small, distant world, a token of our sounds, our science, our images, our music, our thoughts and our feelings. We are attempting to survive our time so we may live into yours.

Sources

Douglas Adams, *The Hitchhiker's Guide to the Galaxy*, with a foreword by Russel T. Davies (Pan Books, London, Basingstoke and Oxford, 2009, first broadcast by BBC Radio 1978, first published 1979)

Niayesh Afshordi, Robert B. Mann and Razieh Pourhasan, 'The Black Hole at the Beginning of Time: do we live in a holographic mirage from another dimension?' *Scientific American*, 311.2 (2014)

Anonymous, 'The Plurality of Worlds', in *Edinburgh Review or Critical Journal*, 102.208 (Longman, Hurst, Rees, Orme, Brown and Green, London, and Adam and Charles Black, Edinburgh, 1855)

Augustine, *Concerning the City of God against the Pagans*, translated by Henry Bettenson (Penguin, Harmondsworth, 1984)

Boethius, *The Consolation of Philosophy*, translated by Victor Watts (Penguin, London and New York, 1999)

Eric Burgess, *Far Encounter: The Neptune System* (Columbia University Press, New York, 1991)

Ritchie Calder, *Man and the Cosmos: The Nature of Science Today* (Penguin, Hardmondsworth, 1968)

Arthur C. Clarke, *The Exploration of Space* (Temple Press, London, 1951)

Arthur C. Clarke, 'The Nine Billion Names of God', in *Of Time and Stars,* Introduction by J.B. Priestley (Penguin in association with Gollanz, Harmondsworth, 1981, first published 1953)

Dante Alighieri, *The Divine Comedy,* translated by Dorothy L. Sayers (Penguin, Harmondsworth: *Hell,* 1949; *Purgatory,* 1955; *Paradise,* 1962)

Thomas Dick, *Celestial Scenery, or the Wonders of the Planetary System Displayed: Illustrating the Perfections of Deity and a Plurality of Worlds, Vol. VII* (Harper & Bros., New York, 1845)

Charles Dickens, *Our Mutual Friend* (Oxford Illustrated Dickens, OUP, London, 1967; first published 1865)

Arthur Conan Doyle, *The Sign of Four*; Introduction by Peter Ackroyd; Notes by Ed Glinert (Penguin Classics, Penguin, London, 2001)

Louis Friedman, *Starsailing: Solar Sails and Interstellar Travel* (Wiley, New York, 1988)

J.B.S. Haldane, *Possible Worlds and Other Essays* (Chatto & Windus, London, 1930)

Jacquetta Hawkes, *A Land* (Pelican, Harmondsworth, 1959; first published 1951)

Stephen Hawking, *A Brief History of Time: From the Big Bang to Black Holes* (Bantam, London, 1988)

Samuel Kinns, *Moses and Geology: or, the Harmony of the Bible with Science* (Cassell, Petter, Galpin & Co., London, 1882)

C.S. Lewis, *The Discarded Image: An Introduction to Medieval and Renaissance Literature* (Cambridge University Press, Cambridge, 1964)

C.S. Lewis, *The Screwtape Letters* (HarperCollins, London, 2001)

C.S. Lewis, *Out of the Silent Planet* (Voyager, London, 2003)

John Milton, *Comus* in 'Selected Poems'; Edited and with an Introduction and Notes by John Leonard (Penguin Classics, Penguin, London, 2007)

Richard Panek, *Seeing and Believing: The Story of the Telescope, or How We Found Our Place in the Universe* (Fourth Estate, London, 2000)

Martin Rees, *Just Six Numbers: The Deep Forces that Shape our Universe* (Weidenfeld & Nicolson, London, 1999)

Leonardo Ricci, 'Dante's insight into Galilean invariance', *Nature*, 434.7034 (2005)

Bertrand Russell, *History of Western Philosophy and its Connection with Political and Social Circumstances from the Earliest Times to the Present Day* (Allen & Unwin, London, 1946)

Carl Sagan, *Pale Blue Dot: A Vision of the Human Future in Space* (Headline, London, 1995)

Jagjit Singh, *Modern Cosmology* (Penguin, Harmondsworth, 1970)

Steven M. Weinberg, *The First Three Minutes: A Modern View of the Origin of the Universe* (Andre Deutsch, London, 1977)

H.G. Wells, *The Wonderful Visit* (J.M. Dent & Sons, The Wayfarer's Library, London, 1914; first published 1895)

H.G. Wells, *The Time Machine* (Book Club Associates, London, 1980; first published 1895)

FURTHER READING

Henry C. Dethloff and Ronald A. Schorn, *Voyager's Grand Tour: To the Outer Planets and Beyond* (Smithsonian Institution Press, Washington, D.C., 2003)

Ben Evans with David M. Harland, *NASA's Voyager Missions: Exploring the Outer Solar System and Beyond* (Springer, London, 2004)

Christopher Riley, Richard Corfield and Philip Dolling, *NASA Voyager 1 & 2: Owners' Workshop Manual* (Haynes Publishing, Yeovil, Somerset, 2015)